THE UNEXPECTED UNIVERSE

BOOKS BY

Loren Eiseley

THE IMMENSE JOURNEY
1957

DARWIN'S CENTURY
1958

THE FIRMAMENT OF TIME
1960

FRANCIS BACON AND THE
MODERN DILEMMA
1962

THE MIND AS NATURE
1962

THE UNEXPECTED UNIVERSE

by

Loren Eiseley

NEW YORK
HARCOURT, BRACE & WORLD, INC.

B.I.70

Library of Congress Catalog Card Number: 67-20308
Printed in the United States of America

To Wolf,
who sleeps forever
with an ice age bone
across his heart,
the last gift
of one
who loved him

ACKNOWLEDGMENTS

I wish to thank the sponsors of the William Haas Lectures of Stanford University, where three of these explorations of the unexpected universe were given, my colleagues in the Institute for Research in the Humanities at the University of Wisconsin, where I was a 1967 guest, and my associates at the Menninger Foundation, where I was a similar visitor. To the John Simon Guggenheim Memorial Foundation and its former director Henry Allen Moe for his patience and his faith I am most grateful. I would like also to express appreciation to the editors of Time-Life Books for permitting the reprinting, with modifications, of a passage from an older article of mine, once confined to a specialized purpose, and to *The American Scholar* and *Life,* in which two of these chapters previously appeared.

LOREN EISELEY

Wynnewood, Pa.
March 3, 1969

CONTENTS

The universe is not only queerer than we suppose, but queerer than we can suppose.

—J. B. S. HALDANE

If you do not expect it, you will not find the unexpected, for it is hard to find and difficult.

—HERACLITUS

THE UNEXPECTED UNIVERSE

ONE

The Ghost Continent

The winds are mad, they know not whence they come, nor whither they would go: and those men are maddest of all that go to sea. —ROBERT BURTON

EVERY man contains within himself a ghost continent—a place circled as warily as Antarctica was circled two hundred years ago by Captain James Cook. If, in addition, the man is a scientist, he will see strange shapes amidst his interior ice floes and be fearful of exposing to the ridicule of his fellows what he has seen. To begin such a personal record it may be well to start with the Odyssean voyages of legend and science. These may defend with something of their own magic the small story of an observer lost upon the fringes of large events. Let it be understood that I claim no discoveries. I claim only the events of a life in science as they were transformed inwardly into something that was whispered to Odysseus long ago.

Like Odysseus, man seeks his spiritual home and is denied it; along his path the shape-shifting immortal monsters of his earlier wanderings assume more sophisticated guises, but they survive because man himself remains and man has called them forth. The almost three-thousand-year-old epic of the Odyssey takes on a particular pertinence today. It possesses a perennial literary freshness that causes it to be translated anew in every generation. It involves an extended journey amid magical obstacles and Cyclopean assailants. Moreover, it can be read as containing the ingredients of both an inward journey of reflection and an outwardly active adventure. Both these journeys threaten to culminate in our time. Man's urge toward space has impelled him to circle the planet, and in the week of December 25, 1968, precisely two hundred years after the navigator James Cook's first great voyage into the Pacific, three American astronauts had returned from the moon. The event lies more than two million years after the first man-ape picked up and used a stone.

Nevertheless, throughout this entire pilgrimage, as reflected in his religious and philosophical thinking, man's technological triumphs have frequently been at odds with his hunger for psychological composure and peace. Thus the epic journey of modern science is a story at once of tremendous achievement, loneliness, and terror. Odysseus' passage through the haunted waters of the eastern Mediterranean symbolizes, at the start of the Western intellectual tradition, the sufferings that the universe and his own nature impose upon homeward-yearning man.

In the restless atmosphere of today all the psychological elements of the Odyssey are present to excess: the driving will toward achievement, the technological cleverness crudely manifest in the blinding of Cyclops, the fierce rejection of the sleepy Lotus Isles, the violence between man and man. Yet, significantly, the ancient hero cries out in desperation, "There is nothing worse for men than wandering."

The words could just as well express the revulsion of a modern thinker over the sight of a nation harried by irrational activists whose rejection of history constitutes an equal, if unrecognized, rejection of any humane or recognizable future. We are a society bemused in its purposes and yet secretly homesick for a lost world of inward tranquillity. The thirst for illimitable knowledge now conflicts directly with the search for a serenity obtainable nowhere upon earth. Knowledge, or at least what the twentieth century acclaims as knowledge, has not led to happiness.

Ours is certainly the most time-conscious generation that has ever lived. Our cameras, our television, our archaeological probings, our C^{14} datings, pollen counts, under-water researches, magnetometer readings have resurrected lost cities, placing them accurately in stratigraphic succession. Each Christmas season the art of ice age Lascaux is placed beside that of Rembrandt on our coffee tables. Views of Pompeii share honors with Chichén Itzá upon the television screen in the living room. We unearth obscure ancestral primates and, in the motion picture "2001," watch a struck fragment of bone fly into the air and become a spaceship drifting among the stars, thus telescoping in an instant the whole technological history of man. We expect the average onlooker to comprehend the symbolism; such a civilization, one must assume, should show a deep veneration for the past.

Strangely, the results are quite otherwise. We appear to be living, instead, amidst a meaningless mosaic of fragments. From ape skull to Mayan temple we contemplate the miscellaneous debris of time like sightseers to whom these mighty fragments, fallen gateways, and sunken galleys convey no present instruction.

In our streets and on our campuses there riots an extremist minority dedicated to the now, to the moment, however absurd, degrading, or irrelevant the moment may be. Such an activism deliberately rejects the past and is determined to start life anew—indeed, to reject the very institutions that feed, clothe, and sustain our swarming millions.

A yearning for a life of noble savagery without the accumulated burdens of history seems in danger of engulfing a whole generation, as it did the French *philosophes* and their eighteenth-century followers. Those individuals who persist in pursuing the mind-destroying drug of constant action have not alone confined themselves to an increasingly chaotic present—they are also, by the deliberate abandonment of their past, destroying the conceptual tools and values that are the means of introducing the rational into the oncoming future.

Their world, therefore, becomes increasingly the violent, unpredictable world of the first men simply because, in losing faith in the past, one is inevitably forsaking all that enables man to be a planning animal. For man's story, in brief, is essentially that of a creature who has abandoned instinct and replaced it with cultural tradition and the hard-won increments of contemplative thought. The lessons of the past have been found to be a reasonably secure instruction for proceeding against the unknown future. To hurl oneself recklessly without method upon a future that we ourselves have complicated is a sheer nihilistic rejection of all that history, including the classical world, can teach us.

Odysseus' erratic journey homeward after the sack of Troy to his own kingdom in Ithaca consumed ten years. There is a sense in which this sea-battered wanderer, who, at one point in concealment, calls himself "Nobody," represents the human journey toward eternity. The sea god Poseidon opposes his passage. He is shipwrecked, escapes monsters, evades the bewitchment of goddesses. In the words of Kazantzakis, he appears to have "a wind chart in his breast for heart."

Yet Odysseus, like Cook and Darwin, the scientific voyagers, is shrewd, self-reliant, and persistent. He is farsighted even when on the magical isles, but he could not always save his companions. They frequently caused him trouble because they were concerned solely with the immediate. Scenting treasure, they opened at the wrong moment the bag that loosed all the winds of the sea upon their vessel. Like man in the mass, they were feckless, unstable, and pursuing the will-o'-the-wisp of the moment. In Homer's words, "they wanted to stay with the Lotus-Eaters and forget the way home."

By contrast, Circe, the great enchantress, says coolly to Odysseus, "There is a mind in you no magic will work on." The remark, however, is two-edged. It is circumspect, as one great lord of thought might speak to another. It hints at the dawn in the Greek mind of that intelligence which we of this age choose to call scientific. Nevertheless, beneath the complimentary words can be sensed a veiled warning. For the man whom no magic will charm may, in the end, find himself, by means of a darker sorcery, upon a shore as desolate as that which Odysseus narrowly escaped in passing the Isle of the Sirens. The Sirens had sung sweetly to him of all knowledge, while about them lay dead men's bones. If living for the day and the senses is the folly of the thoughtless, so also is there danger in that insatiable hunger for power which besets the human intellect. Far more than modern men, Homer is wary of that vaulting pride the Greeks called *hubris,* which is an affront to the immortal gods.

It was once said, half in irony, by an ancient geographer that "you will find where Odysseus wandered when you find the cobbler who stitched the bag of the winds." Doubtless this is true, but is not man similarly the product of such an untraceable cobbler? And, even more, is not each individual life a bag full of surging dreams and compulsions imprisoned in human skin by that same cobbler, and equally capable of inadvertent release? Yet for all the vagaries of human voyaging amidst inward and outward tempests, the mariners of three thousand years ago had begun scientifically to watch the stars. Homer himself was acquainted with the guide stars of the seafarers. We know from the *Odyssey* that the constellation of the Great Bear was never wetted by the waves on its night circling above the Mediterranean.

If one now turns to the Odyssean voyages of science in the eighteenth and nineteenth centuries, one will come surprisingly close both to that shadowy Cimmerian land where the sun is hidden and to that rift in time where life leaped backward to confront itself upon its evolutionary road. The voyages that produced these observations were as irretraceable and marvel-filled as any upon the lost sea charts of Odysseus.

2

It has been said of Captain James Cook that no discoverer ever measured his claims with more moderation. Yet among the great mariners none ran greater risks for the purposes of close inshore mapping. None sailed farther, or sailed under weightier secret instructions. Masterful and solitary, but with a superb gift of leadership, he endured the vanities of his scientific associates as he endured plotting natives and the hard diet of the sea.

He was a man supremely indifferent to every circumstance, who brought to the tenth year of his Pacific voyaging the same ingenuity and doggedness that had brought Odysseus home from Troy. Like Odysseus he could practice wise restraint; like Odysseus he could improvise against the future. Unlike the more primitive warriors of the Greek bronze age, however, he was not vengeful. He saw in the looming future the possibilities offered by Australia for civilized settlement, and directed attention to the furtherance of such potentials. So vast was the range of his wanderings and so attractive the Pacific isles he visited and explored that they have veiled from our memory the storms and darkness that hovered over his greatest achievement, the circumnavigation of Antarctica.

Men of today frequently turn to science for such knowledge of the hazardous future as can be gained by mortals. In Homer's time it was believed that this information might be sought among the dead. At Circe's instigation Odysseus reached toward the world's edge the mist-shrouded land of Hades, the dwelling place of the shades:

> hidden in fog and closed, nor does Helios, the radiant
> sun, ever break through the dark . . .
> always a drear night is spread.

It was there that the dead Theban, Tiresias, foretold the end of Odysseus' voyage, further remarking that in old age "death will come to you from the sea, in some altogether unwarlike way."

Cook, who opened the Pacific frontier to science, is largely associated in the public mind with those Lotus Isles of forgetfulness that lie in Polynesia. In reality, however, he had received, like Odysseus, assignment to a more desperate errand. The adventure was as stark in his day as the adventure into space is in ours. In terms of supporting equipment, it was equally, if not more, dangerous, for Cook's mission was to penetrate the unknown region at the bottom of the globe. In 1768 his public orders were ostensibly to proceed to the Pacific and observe from Tahiti the transit of Venus. On the island he was to open his sealed instructions, which read in part:

> Whereas there is reason to imagine that a Continent or Land of great extent, may be found to . . . the Southward of the Track of any former Navigators . . . you are to proceed to the southward in order to make discovery of the Continent above mentioned. . . .

Whence had emerged this conception of a continent unknown but supposed to be clothed with verdure and inhabited by living people whose wares and activities might be of interest to the Crown? Since the days of Ptolemy a great southern land mass, a continent labeled Terra Incognita, had floated down through the centuries on a succession of maps. Although belief in such a continent wavered and died down, certain sightings of islands in the sixteenth century restimulated the hopes of geographers. Visions were entertained of a rich and habitable continent south and westward of South America. It moved upon the eighteenth-century sea charts as elusively as Melville's great white whale, in all meridians.

By Cook's time, in the late eighteenth century, an ambitious scholar-merchant, Alexander Dalrymple, whose hobby was the memoirs of the early voyagers, had become a believer in the ghost continent. Dalrymple thought this huge mainland must be necessary to balance the earth on its axis and that its human population must be numbered in the millions. Dalrymple wished to lead an expedition there to establish trade relations. Cook, a proven naval commander with long mapping and coastal sailing experience, was chosen instead. Dalrymple was bitterly disappointed and harassed Cook after his first voyage of 1768, saying, "I would not have come back in ignorance." Cook had stated, after long voyaging, "I do not believe any such thing exists, unless [and here he proved prophetic] in a high latitude." Dalrymple succeeded in creating such doubt and confusion that the Admiralty decided on a second expedition, but again Cook was the chosen officer; the ghost continent was once more pursued through the shifting degrees of latitude.

Antarctica is another world. Instead of discovering a living continent, Cook, like Odysseus, came to a land of Cimmerian darkness. Huge icicles hung on the ship's sails and rigging. The pack ice "exhibited such a variety of figures that there was not an animal on earth that was not represented by it." A breeding sow on board farrowed nine pigs, every one of which was killed by cold in spite of attempts to save them. Scurvy appeared. In the after cabin a gentleman died. A sailor dropped from the rigging and vanished beneath the ice floes. Cavernous icebergs, against which the waves resounded, inspired exclamations of admiration and horror. Vast-winged gray albatrosses drifted by in utter silence.

In 1773 Cook first crossed the Antarctic Circle,

where, in the words of one of his scientist passengers, "We were . . . wrapped in thick fogs, beaten with showers of rain, sleet, hail and snow . . . and daily ran the risk of being shipwrecked." Cook himself, after four separate and widely removed plunges across the Circle, speaks, as does Homer, of lands "never to yield to the warmth of the sun." His description of an "inexpressibly horrid Antarctica" resounds like an Odyssean line. Terra Incognita Australis had been circumnavigated at last, its population reduced to penguins. If there was land at all beyond the ice barrier, it was the frozen world of another planet. Only the oaths of sailors splintered and re-echoed amidst the pinnacles of ice.

"I can be bold to say," pronounced Cook, turning north toward the summer isles, "that no man will venture farther than I have done." In eighteenth-century terms he was right, just as now the way beyond the moon lies dark and inexpressibly desolate and costly for men to follow. The mariners in their sea jackets and canvas headgear were not equipped for the long land traverse that, in the twentieth century, took Scott and Amundsen to the pole; even then the fierce ice gods took away one man's homecoming. With grim amusement Cook heard his officers suggest that if Terra Incognita existed it lay north, in sunnier climes. Ignoring them, perhaps with Dalrymple in mind, he tried one more run southward before he swung away. The highborn scientists were frequently to complain uneasily that he never told them where he was going. What was he expected to do, sailing under secret orders in the wastes below the Circle? He might, if he had chosen, have responded like Odysseus, "I am a man. I am not a god."

A little over a half century after Cook's death in Hawaii, an inexperienced voyager into the bottomless sea of time came to anchor in the Galápagos. It was the young Charles Darwin, fresh from his wanderings on the pampas and his Andean ascents. At Tierra del Fuego he had glimpsed from the *Beagle* the stormy, racing waters through which Cook's ships, the *Endeavor* and the *Resolution,* had ridden on their long world journeys. Now the *Beagle* had rung down its anchor at what the Spanish had called, with singular discernment, the "Encantadas," the Enchanted Isles.

Odysseus had come similarly upon Circe's island, only to find his crew transformed into animals—specifically, into pigs. When, at his behest, the changelings were created men once more, they took on a more lively and youthful appearance. By the sixteenth century the Florentine writer Giovanni Battista Gelli had produced his *Circe,* in which a variety of animals refused Odysseus' offer to restore them to their original form. Their arguments for remaining as they were constitute an ingenious commentary on the human condition. From rabbit to lion the animals are united in being done with humanity. Not all the argumentative wiles of Odysseus can talk them back into the shape of *Homo sapiens*. The single exception proves to be a dubious Greek philosopher immured in the body of an elephant. He alone consents to a renewed transformation.

The Encantadas are the means by which the whole Circean labyrinth of organic change was precipitated upon the mind of man. What had appeared to Odysseus as the trick of a goddess was, in actuality, the shape shifting of the incomprehensible universe itself. Among the upthrust volcanic chimneys, which Darwin compared to "the cultivated parts of the Infernal Regions," the young naturalist meditated upon a flora and fauna seemingly distinct, as in the case of the famed Galápagos tortoises, from the living inhabitants of the continents.

Circe kept herself hidden, but it was evident to the wondering Darwin that there was a power hidden in time and isolation that alone could transmute, not just men, but all things living, into wavering shadows. He had entered by the mysterious doorway of the Galápagos into a sea as vast in its own way as the limitless Pacific. Yet even here upon the ocean of time the ghostly sail of Cook's vessel passes by. Cook's surgeon-naturalist, William Anderson, records in his journal of the third and fatal voyage—that voyage which Cook himself had preternaturally referred to as the "last"—that animals and human beings must be attributed to the different stocks "from which they sprung before their arrival in the south sea, or we must believe that at the creation every particular island was furnished with its inhabitants in the same manner as with its peculiar plants." Anderson's words do not precisely express evolutionary ideas, but they hint prophetically at the puzzling thoughts—to a degree, heretical—that had arisen and would not be downed about the role of islands in the biology of the living world.

The momentary phantom of the *Resolution* fades on the infinite waters of the past. William Anderson is dead on the Bering passage, but others will look and ask, and ask again, until the answer arises upon Darwin's return to London—that London which, like Odysseus searching the horizon for Ithaca, he yearns for in a note to his former teacher John Henslow: "Oh, to be home again without one single novel object near me." But the novel objects are there, fixed in the memory, ever to be relived. Each covert of the Enchanted Archipelago will forever resound with the reptilian hisses of antediluvian antiquity.

The memory of the modern scholar will, I suspect, be wrenched by the thought of all that Captain Cook patiently endured as well as learned from Sir Joseph Banks, the well-intentioned, blustering, aristocratic naturalist, and his sometimes complaining colleagues of the first and second voyages. Only once did the restrained Captain allow himself the liberty of a quarter-deck comment, involving the violation of the commander's printing rights by the naturalist Forster and his son. The careless assignment to Cook by the scribe Hawkesworth of some undiplomatic remarks derived from Banks's private journal did nothing to allay the Captain's wrath. "Damn and blast all scientists," he exploded to Lieutenant King on the eve of the third voyage.

Cook had reason for his spleen. Yet it is worth recording that this son of a Yorkshire laborer left a record of his voyages far more accurate than the first version produced by Hawkesworth, who had been assigned by the Admiralty to prettify Cook's work and make it palatable. In modern terms, Cook was, on the whole, a magnificent and tolerant anthropologist who, at every inhabited island, had to improvise his Odyssean role. It was not scientists who died on the beach of Kealakekua in Hawaii. Instead, it was the underrated circumnavigator of Antarctica, the genuine Ahab of the ghost continent. Cook had, times without number, brought his ships off dangerous shoals and, in addition, penetrated the high latitudes of both poles.

We inhabitants of a scientific era may well prize our accomplishments, but it behooves us to acknowledge that without the skills of the relentless Odyssean voyager, Joseph Banks and all his colleagues might well have fed the coral upon some tropic reef. Banks is believed to have anonymously paid Cook a just, belated tribute in the *Morning Chronicle* when the news of his death was announced in London. "Cook's competence changed the face of the world" was to be a mature, twentieth-century judgment.

James Cook himself would have cast a proud, cold eye upon land-bound opinion. He had emerged unheralded and alone from the grimy foreshore of history. The horizon, the pack ice, and the albatross would have been all he asked as memorials of his passage—these and the carefully drawn maps for those who followed. It may be that the space captains, when and if they leave the solar system, will alone come to understand that remote, serene indifference.

There is, however, one thing more to be considered: the Odyssean journey in which the mind turns homeward, seeking surcease from outer triumphs. Perhaps this crossing of the two contradictory impulses in the mind of man is nowhere better expressed than by the Italian poet Giovanni Pascoli in "Ultimo Viaggio" ("The Last Voyage"), published in 1904. Pascoli realized that Odysseus' return to Ithaca, his homeward goal, was in a sense an anticlimax—that the magical spell wrought by Circe would follow the hero into the prosaic world.

Pascoli picks up the Odyssean tale when Odysseus, grown old and restless, drawn on by migratory birds, sets forth to retrace his magical journey, the journey of all men down the pathway of their youth, the road beyond retracing. Circe's isle lies at last before the wanderer in the plain colors of reality. Circe and whatever she represents have vanished. Much as Darwin might have viewed the Galápagos in old age, Odysseus passes the scenes of the marvelous voyage with all the obstacles reduced to trifles. The nostalgia of space, which is what the Greeks meant by nostalgia, that is, the hunger for home, is transmuted by Pascoli into the hunger for lost time, for the forever vanished days. The Sirens no longer sing, but Pascoli's Odysseus, having made his inward journey, understands them. Knowledge without sympathetic perception is barren. Odysseus in his death is carried by the waves to Calypso, who hides him in her hair. "Nobody" has come home to Nothingness.

3

Those archaeologists and students of folklore who frequent the Scandinavian swamps in which bog acids have preserved such bodies as those of Tollund man—men who were alive upon earth in a time almost coeval with Homer—speak of a strange crossroads religion. An earth goddess was driven in a massive coach through gloomy northern forests. She was peripatetic and did not stay, but the ox-drawn vehicle in which she lumbered over abandoned and ancient roads paused, with curtains drawn, at ignorant villages.

Out of the rising fog wisps shambled her priests and obedient worshipers. No one was allowed to thrust his head behind the draped curtains of the ornamented cart or to query the immobile coachman or to touch with worshipful fingers the steaming oxen in the night. After the proper ritual, a selected human sacrifice would be cast down into the waiting bog. The coach would lumber again into the night, across wild heaths and forbidden pathways.

But suppose—the thought strikes me suddenly as I startle awake at midnight, hearing cries and rumblings in my hotel corridors—suppose the wheels of the great car are still revolving before its attendant worshipers. Suppose further—and I sit up at this and shiver uncontrollably as I hear the mead-soaked voices and the running steps—suppose the same awkward coach still lurches through the darker hours of our assembled scientific priesthood.

Suppose that in the ancient car there sits in one age the masked face of Newton, his world machine ticking like a remorseless clock in the dead and confined air; or suppose that Darwin lurks concealed behind the curtains, and all is wild uncertainty and change in the misty features of his company; or that Doctor Freud looks coldly and contemplatively down upon a sea of fleering goblin faces. Or is it the Abbé Lemaître's followers who hear the alternate expansion and contraction of nature's pounding heart, like a rhythmic drum amidst the receding coals of the night stars? Or imagine that, silhouetted gigantically in the fierce rays of atomic light streaming from the carriage, four sinister horsemen trample impatiently, while within a muffled voice cries out to the assembled masses: "God is dead. All is permitted."

Suppose, I think, lying awake and weary in my blankets, the figure in the coach is a changeling and its true face is no face, as Odysseus was "no one" until he shouted a vengeful name before the Cyclops. Or deduce that behind the concealing drapery, hooded in a faceless cowl, there is caught only the swirling vapor of an untamed void whose vassals we are—we who fancy ourselves as the priesthood of powers safely contained and to be exhibited as evidences of our own usurping godhood.

I hear now, as I have heard before, the far-off, long, premonitory trundling of the dreadful cart at midnight. I think of the face of No Face—protean, dissembling, alternating death's-head and beautiful Circean goddess, forever forbidden to descend and touch common flesh. I think of her acolytes, ourselves, toiling in a hundred laboratories with our secret visions of what is, or may not be, while the wild reality always eludes our grasp.

Long ago a Greek named Plato, who was also a voyager not unfamiliar with Odyssean wanderings, remarked after much thought, "We must take the best and most irrefragable of human doctrines and embark on that, as if it were a raft on which to risk the voyage of life." Plato makes, however, one humble addition to his meditation, one that marks the Greek distaste of overvaulting pride. "Unless," he adds wistfully, "it were possible to find a stronger vessel, some Divine Word on which we might take our journey more surely and with confidence."

Earlier I have dwelt upon the magical incidents that beset Odysseus on his voyage. These are the stories remembered through unnumbered generations, of the island of the Cyclops, of Circe, of Calypso, of the cattle of the sun, of the Lotus-Eaters and the Sirens, of the dreadful bag of the four winds inadvertently loosed in mid-passage. The Odyssey is, as the critic Howard Clarke once observed, the world of the folk tale. It is the arena of uncertain violence that confronts man on the voyage of life—sex, irresponsibility, inordinate hungers that evoke equally monstrous figures from the human subconscious. Such a world is like the Jack-and-the-Beanstalk world of childhood, whose memories linger more clearly than many of the events of later life.

Once it is raised, not only is it difficult to subdue such a world again, but anything that follows it is apt to prove an anticlimax. Odysseus' goal is home, but after the adventures on the mysterious sea where Proteus can be found basking on a rock among his creatures, the suitors and their puerile human rapacity are singularly unattractive. Odysseus himself loses stature when he is reduced to lynching helpless if in-obedient women servants. Something vanishes from the tale at this point, no matter how necessary it was for Homer to complete the Odyssean voyage. The poet has been sufficiently perceptive, however, to lay upon Odysseus, through the device of prophecy, the injunction of a further journey.

The poet, or, rather, his hero, has returned in middle age to a household that will obviously not long give him scope to breathe. Odysseus must, in other words, be given a legitimate reason to escape from Ithaca. This has been so strongly felt by generations of readers that from Dante through Tennyson to Kazantzakis poets have been impelled to launch the immortal navigator once more into the realms of dream. Perhaps among these poets Pascoli alone was wise enough to visualize an end in which the trivial and magicless themselves are transmuted by human wisdom into a timeless dimension having its own enchanted reality.

In this connection, there is one event involving Odysseus' homecoming, one episode that, in the mundane world of Ithaca, shines like a far light reflected from Calypso's isle. Oddly enough, that other voyager on more recent seas, Charles Darwin, experienced and recorded a similar episode after his five-year absence from Shrewsbury. Odysseus' great dog Argos, abused and cast out on a dung heap, wagged his tail and was the first to recognize his old master after nineteen years of absence. In a like way Darwin's favorite dog recognized him after his five-year absence on the *Beagle*.

It is not necessary to cavil over the great age of Odysseus' dog. The surprising thing about the story's descent through the millennia is that it comes from a fierce and violent era, yet it bespeaks some recognized bond between man and beast. The tie runs beyond the cities into some remote glade in a far forest where man willingly accepted the help of his animal kin. Though men in the mass forget the origins of their need, they still bring wolfhounds into city apartments, where dog and man both sit brooding in wistful discomfort.

The magic that gleams an instant between Argos and Odysseus is both the recognition of diversity and the need for affection across the illusions of form. It is nature's cry to homeless, far-wandering, insatiable man: "Do not forget your brethren, nor the green wood from which you sprang. To do so is to invite disaster."

Many great writers have written private meanings into their versions of the Odyssean voyage. In the twentieth century particularly, Odysseus has been seen as a symbol of the knowledge-hungry scientist, the Faustian penetrator of space and time. But, as scientists, we have sometimes forgotten the inward journey so poetically expressed by Pascoli in "The Last Voyage," that inward journey whose true meaning was long ago expressed by Circe's cryptic warning. "Magic cannot touch you," she had said to Odysseus, but today we know that the heart untouched by the magic of wonder may come to an impoverished age. Cook, the navigator, died in the Lotus Isles, perhaps fortunate that he never returned to his Penelope. Darwin, who voyaged as deeply into time, came home to Down, where he is said to have slept ill. In fact, it has been reported on good authority that he walked so late he met the foxes trotting home at dawn.

Thus, in the heart of man, and, above all, within this turbulent century, the Odyssean voyage stands as a symbol of both man's homelessness and his power, a power more unregenerate than that which drove Odysseus to string the great bow before the suitors. Long ago, when the time to Homer might still be numbered in centuries, Plotinus wrote of the soul's journey, "It shall come, not to another, but to itself." It is possible to add that for the soul to come to its true self it needs the help and recognition of the dog Argos. It craves that empathy clinging between man and beast, that nagging shadow of remembrance which, try as we may to deny it, asserts our unity with life and does more. Paradoxically, it establishes, in the end, our own humanity. One does not meet oneself until one catches the reflection from an eye other than human.

It has been asserted that we are destined to know the dark beyond the stars before we comprehend the nature of our own journey. This may be true. But we know also that our inward destination lies somewhere a long way past the reef of the Sirens, who sang of knowledge but not of wisdom. Beyond that point, if perchance we reach it, exists the realm of Plato's undiscovered word. It is a divine word, or so Plato gropingly called it, for he hoped against hope that it might suffice to guide our human pilgrimage when, in the ritual language of the *Odyssey,* "the sun has set and all the journeying ways are darkened."

In our time, however, the mind still persists in traveling along those darkened sea paths where all manner of strange creatures swarm. For myself, I have penetrated as far as I could dare among rain-dimmed crags and seascapes. But there is more, assuredly there is still more, as Circe tried to tell Odysseus when she warned that death would come to him from the sea. She meant, I think now, the upwelling of that inner tide which finally engulfs each traveler.

I have listened belatedly to the warning of the great enchantress. I have cast, while there was yet time, my own oracles on the sun-washed deck. My attempt to read the results contains elements of autobiography. I set it down just as the surge begins to lift, towering and relentless, against the reefs of age.

TWO

The Unexpected Universe

Imagine God, as the Poet saith, Ludere in Humanis, *to play but a game at Chesse with this world; to sport Himself with making little things great, and great things nothing; Imagine God to be at play with us, but a gamester. . . .* —JOHN DONNE

A British essayist of distinction, H. J. Massingham, once remarked perceptively that woods nowadays are haunted not by ghosts, but by a silence and man-made desolation that might well take terrifying material forms. There is nothing like a stalled train in a marsh to promote such reflections—particularly if one has been transported just beyond the environs of a great city and set down in some nether world that seems to partake both of nature before man came and of the residue of what will exist after him. It was night when my train halted, but a kind of flame-wreathed landscape attended by shadowy figures could be glimpsed from the window.

After a time, with a companion, I descended and

strolled forward to explore this curious region. It turned out to be a perpetually burning city dump, contributing its miasmas and choking vapors to the murky sky above the city. Amidst the tended flames of this inferno I approached one of the grimy attendants who was forking over the rubbish. In the background, other shadows, official and unofficial, were similarly engaged. For a moment I had the insubstantial feeling that must exist on the borders of hell, where everything, wavering among heat waves, is transported to another dimension. One could imagine ragged and distorted souls grubbed over by scavengers for what might usefully survive.

I stood in silence watching this great burning. Sodden papers were being forked into the flames, and after a while it crossed my mind that this was perhaps the place where last year's lace valentines had gone, along with old Christmas trees, and the beds I had slept on in childhood.

"I suppose you get everything here," I ventured to the grimy attendant.

He nodded indifferently and drew a heavy glove across his face. His eyes were red-rimmed from the fire. Perhaps they were red anyhow.

"Know what?" He swept a hand outward toward the flames.

"No," I confessed.

"Babies," he growled in my ear. "Even dead babies sometimes turn up. From there." He gestured contemptuously toward the city and hoisted an indistinguishable mass upon his fork. I stepped back from the flare of light, but it was only part of an old radio cabinet. Out of it had once come voices and music and laughter, perhaps from the twenties. And where were the voices traveling now? I looked at the dangling fragments of wire. They reminded me of something, but the engine bell sounded before I could remember.

I made a parting gesture. Around me in the gloom dark shapes worked ceaselessly at the dampened fires. My eyes were growing accustomed to their light.

"We get it all," the dump philosopher repeated. "Just give it time to travel, we get it all."

"Be seeing you," I said irrelevantly. "Good luck."

Back in my train seat, I remembered unwillingly the flames and the dangling wire. It had something to do with an air crash years ago and the identification of the dead. Anthropologists get strange assignments. I put the matter out of my mind, as I always do, but I dozed and it came back: the box with the dangling wires. I had once fitted a seared and broken skullcap over a dead man's brains, and I had thought, peering into the scorched and mangled skull vault, it is like a beautiful, irreparably broken machine, like something consciously made to be used, and now where are the voices and the music?

"We get it all," a dark figure said in my dreams. I sighed, and the figure in the murk faded into the clicking of the wheels.

One can think just so much, but the archaeologist is awake to memories of the dead cultures sleeping around us, to our destiny, and to the nature of the universe we profess to inhabit. I would speak of these things not as a wise man, with scientific certitude, but from a place outside, in the role, shall we say, of a city-dump philosopher. Nor is this a strained figure of speech. The archaeologist is the last grubber among things mortal. He puts not men, but civilizations, to bed, and passes upon them final judgments. He finds, if imprinted upon clay, both our grocery bills and the hymns to our gods. Or he uncovers, as I once did in a mountain cavern, the skeleton of a cradled child, supplied, in the pathos of our mortality, with the carefully "killed" tools whose shadowy counterparts were intended to serve a tiny infant through the vicissitudes it would encounter beyond the dark curtain of death. Infinite care had been lavished upon objects that did not equate with the child's ability to use them in this life. Was his spirit expected to grow to manhood, or had this final projection of bereaved parental care thrust into the night, in desperate anxiety, all that an impoverished and simple culture could provide where human affection could not follow?

In a comparable but more abstract way, the modern mind, the scientific mind, concerned as it is with the imponderable mysteries of existence, has sought to equip oncoming generations with certain mental weapons against the terrors of ignorance. Protectively, as in the case of the dead child bundled in a cave, science has proclaimed a universe whose laws are open to discovery and, above all, it has sought, in the words of one of its greatest exponents, Francis Bacon, "not to imagine or suppose, but to *discover* what nature does or may be made to do."

To discover what nature does, however, two primary

restrictions are laid upon a finite creature: he must extrapolate his laws from what exists in his or his society's moment of time and, in addition, he is limited by what his senses can tell him of the surrounding world. Later, technology may provide the extension of those senses, as in the case of the microscope and telescope. Nevertheless the same eye or ear with which we are naturally endowed must, in the end, interpret the data derived from such extensions of sight or hearing. Moreover, science since the thirteenth century has clung to the dictum of William of Ockham that hypotheses must not be multiplied excessively; that the world, in essence, is always simple, not complicated, and its secrets accessible to men of astute and sufficiently penetrating intellect. Ironically, in the time of our greatest intellectual and technological triumphs one is forced to say that Ockham's long-honored precepts, however well they have served man, are, from another view, merely a more sophisticated projection of man's desire for order—and for the ability to control, understand, and manipulate his world.

All of these intentions are commendable enough, but perhaps we would approach them more humbly and within a greater frame of reference if we were to recognize what Massingham sensed as lying latent in his wood, or what John Donne implied more than three centuries ago when he wrote:

> I am rebegot
> of absence, darknesse, death:
> Things which are not.

Donne had recognized that behind visible nature lurks an invisible and procreant void from whose incomprehensible magnitude we can only recoil. That void has haunted me ever since I handled the shattered calvarium that a few hours before had contained, in microcosmic dimensions, a similar lurking potency.

Some years previously, I had written a little book of essays in which I had narrated how time had become natural in our thinking, and I had gone on to speak likewise of life and man. In the end, however, I had been forced to ask, How Natural is Natural?—a subject that raised the hackles of some of my scientifically inclined colleagues, who confused the achievements of their disciplines with certitude on a cosmic scale. My very question thus implied an ill-concealed heresy. That heresy it is my intent to pursue further. It will involve us, not in the denigration of science, but, rather, in a farther stretch of the imagination as we approach those distant and wooded boundaries of thought where, in the words of the old fairy tale, the fox and the hare say good night to each other. It is here that predictability ceases and the unimaginable begins—or, as a final heretical suspicion, we might ask ourselves whether our own little planetary fragment of the cosmos has all along concealed a mocking refusal to comply totally with human conceptions of order and secure prediction.

The world contains, for all its seeming regularity, a series of surprises resembling those that in childhood terrorized us by erupting on springs from closed boxes. The world of primitive man is not dissimilar. Lightning leaps from clouds, something invisible rumbles in the air, the living body, spilling its mysterious red fluid, lies down in a sleep from which it cannot waken. There are night cries in the forest, talking waters, guiding omens, or portents in the fall of a leaf. No longer, as with the animal, can the world be accepted as given. It has to be perceived and consciously thought about, abstracted, and considered. The moment one does so, one is outside of the natural; objects are each one surrounded with an aura radiating meaning to man alone. To a universe already suspected of being woven together by unseen forces, man brings the organizing power of primitive magic. The manikin that is believed to control the macrocosm by some sympathetic connection is already obscurely present in the poppet thrust full of needles by the witch. Crude and imperfect, magic is still man's first conscious abstraction from nature, his first attempt to link disparate objects by some unseen attraction between them.

2

If we now descend into the early years of modern science, we find the world of the late eighteenth and early nineteenth centuries basking comfortably in the conception of the balanced world machine. Newton had established what appeared to be the reign of universal order in the heavens. The planets—indeed, the whole cosmic engine—were self-regulatory. This passion for order controlled by a divinity too vast to be concerned with petty miracle was slowly extended to earth. James Hutton glimpsed, in the long erosion and renewal of the continents by subterranean uplift, a similar "beautiful machine" so arranged that recourse to the "preternatural," or "destructive accident," such as the Mosaic Deluge, was unnecessary to account for the physical features of the planet.

Time had lengthened, and through those eons, law, not chaos, reigned. The imprint of fossil raindrops similar to those of today had been discovered upon ancient shores. The marks of fossil ripples were also observable in uncovered strata, and buried trees had absorbed the sunlight of far millennia. The remote past was one with the present and, over all, a lawful similarity decreed by a Christian Deity prevailed.

In the animal world, a similar web of organization was believed to exist, save by a few hesitant thinkers. The balanced Newtonian clockwork of the heavens had been transferred to earth and, for a few decades, was destined to prevail in the world of life. Plants and animals would be frozen into their existing shapes; they would compete but not change, for change in this system was basically a denial of law. Hutton's world renewed itself in cycles, just as the oscillations observable in the heavens were similarly self-regulatory.

Time was thus law-abiding. It contained no novelty and was self-correcting. It was, as we have indicated, a manifestation of divine law. That law was a comfort to man. The restive world of life fell under the same dominion as the equally restive particles of earth. Organisms oscillated within severely fixed limits. The smallest animalcule in a hay infusion carried a message for man; the joints of an insect assured him of divine attention. "In every nature and every portion of nature which we can descry," wrote William Paley in a book characteristic of the period, "we find attention bestowed upon even the minutest parts. The hinges in the wing of an earwig . . . are as highly wrought as if the creator had nothing else to finish. We see no signs of diminution of care by multiplicity of objects, or distraction of thought by variety. We have no reason to fear, therefore, our being forgotten, or overlooked, or neglected." Written into these lines in scientific guise is the same humanly protective gesture that long ago had heaped skin blankets, bone needles, and a carved stick for killing rabbits into the burial chamber of a child.

This undeviating balance in which life was locked was called "natural government" by the great anatomist John Hunter. It was, in a sense, like the cyclic but undeviating life of the planet earth itself. That vast elemental creature felt the fall of raindrops on its ragged flanks, was troubled by the drift of autumn leaves or the erosive work of wind throughout eternity. Nevertheless, the accounts of nature were strictly kept. If a continent was depressed at one point, its equivalent arose elsewhere. Whether the item in the scale was the weight of a raindrop or a dislodged boulder on a mountainside, a dynamic balance kept the great beast young and flourishing upon its course.

And as it was with earth, so also with its inhabitants.

"There is an equilibrium kept up among the animals by themselves," Hunter went on to contend. They kept their own numbers pruned and in proportion. Expansion was always kept within bounds. The struggle for existence was recognized before Darwin, but only as the indefinite sway of a returning pendulum. Life was selected, but it was selected for but one purpose: vigor and consistency in appearance. The mutative variant was struck down. What had been was; what would be already existed. As in the case of that great animal the earth, of the living flora and fauna it could be said that there was to be found "no vestige of a beginning, —no prospect of an end." An elemental order lay across granite, sea, and shore. Each individual animal peered from age to age out of the same unyielding sockets of bone. Out of no other casements could he stare; the dweller within would see leaf and bird eternally the same. This was the scientific doctrine known as uniformitarianism. It had abolished magic as it had abolished the many changes and shape shiftings of witch doctors and medieval necromancers. At last the world was genuinely sane under a beneficent Deity. Then came Darwin.

3

At first, he was hailed as another Newton who had discovered the laws of life. It was true that what had once been deemed independent creations—the shells in the collector's cabinet, the flowers pressed into memory books—were now, as in the abandoned magic of the ancient past, once more joined by invisible threads of sympathy and netted together by a common ancestry. The world seemed even more understandable, more natural than natural. The fortuitous had become fashionable, and the other face of "natural government" turned out to be creation. Life's pendulum of balance was an illusion.

Behind the staid face of that nature we had worshiped for so long we were unseen shapeshifters. Viewed in the long light of limitless time, we were optical illusions whose very identity was difficult to fix. Still, there was much talk of progress and perfection. Only later did we begin to realize that what Charles Darwin had introduced into nature was not Newtonian predictability but absolute random novelty. Life was bent, in the phrase of Alfred Russel Wallace, upon "indefinite departure." No living thing, not even man, understood upon what journey he had embarked. Time was no longer cyclic or monotonously repetitious.* It was historic, novel, and unreturning. Since that momentous discovery, man has, whether or not he realizes or accepts his fate, been moving in a world of contingent forms.

Even in the supposedly stable universe of matter, as it was viewed by nineteenth-century scientists, new problems constantly appear. The discovery by phys-

*For purposes of space I have chosen to ignore the short-lived doctrine of the early century known as catastrophism, since I have treated it at length elsewhere.

icists of antimatter particles having electric charges opposite to those that compose our world and unable to exist in concert with known matter raises the question of whether, after all, our corner of the universe is representative of the entire potentialities that may exist elsewhere. The existence of antimatter is unaccounted for in present theories of the universe, and such peculiarities as the primordial atom and the recently reported flash of the explosion at the birth of the universe, as recorded in the radio spectrum, lead on into unknown paths.

If it were not for the fact that familiarity leads to assumed knowledge, we would have to admit that the earth's atmosphere of oxygen appears to be the product of a biological invention, photosynthesis, another random event that took place in Archeozoic times. That single "invention," for such it was, determined the entire nature of life on this planet, and there is no possibility at present of calling it preordained. Similarly, the stepped-up manipulation of chance, in the shape of both mutation and recombination of genetic factors, which is one result of the sexual mechanism, would have been unprophesiable.

The brain of man, that strange gray iceberg of conscious and unconscious life, was similarly unpredictable until its appearance. A comparatively short lapse of geological time has evolved a humanity that, beginning in considerable physical diversity, has increasingly converged toward a universal biological similarity, marked only by a lingering and insignificant racial differentiation. With the rise of *Homo sapiens* and the final perfection of the human brain as a manipulator of symbolic thought, the spectrum of man's possible social behavior has widened enormously. What is essentially the same brain biologically can continue to exist in the simple ecological balance of the Stone Age or, on the other hand, may produce those enormous inflorescences known as civilizations. These growths seemingly operate under their own laws and take distinct and irreversible pathways. In an analogous way, organisms mutate and diverge through adaptive radiation from one or a few original forms.

In the domain of culture, man's augmented ability to manipulate abstract ideas and to draw in this fashion enormous latent stores of energy from his brain has led to an intriguing situation: the range of his *possible* behavior is greater and more contradictory than that which can be contained within the compass of a single society, whether tribal or advanced. Thus, as man's penetration into the metaphysical and abstract has succeeded, so has his capacity to follow, in the same physical body, a series of tangential roads into the future. Likeness in body has, paradoxically, led to diversity in thought. Thought, in turn, involves such vast institutional involutions as the rise of modern science, with its intensified hold upon modern society.

All past civilizations of men have been localized and have had, therefore, the divergent mutative quality to which we have referred. They have offered choices to men. Ideas have been exchanged, along with technological innovations, but never on so vast, overwhelming, and single-directed a scale as in the present. Increasingly, there is but one way into the future: the technological way. The frightening aspect of this situation lies in the constriction of human choice. Western technology has released irrevocable forces, and the "one world" that has been talked about so glibly is frequently a distraught conformity produced by the centripetal forces of Western society. So great is its power over men that any other solution, any other philosophy, is silenced. Men, unknowingly, and whether for good or ill, appear to be making their last decisions about human destiny. To pursue the biological analogy, it is as though, instead of many adaptive organisms, a single gigantic animal embodied the only organic future of the world.

4

Archaeology is the science of man's evening, not of his midday triumphs. I have spoken of my visit to a flame-wreathed marsh at nightfall. All in it had been substance, matter, trailing wires and old sandwich wrappings, broken toys and iron bedsteads. Yet there was nothing present that science could not reduce into its elements, nothing that was not the product of the urban world whose far-off towers had risen gleaming in the dusk beyond the marsh. There on the city dump had lain the shabby debris of life: the waxen fragment of an old record that had stolen a human heart, wilted flowers among smashed beer cans, the castaway knife of a murderer, along with a broken tablespoon. It was all a maze of invisible, floating connections, and would be until the last man perished. These forlorn materials had all been subjected to the dissolving power of the human mind. They had been wrenched from deep veins of rock, boiled in great crucibles, and carried miles from their origins. They had assumed shapes that, though material enough, had existed first as blueprints in the profound darkness of a living brain. They had been defined before their existence, named and given shape in the puff of air that we call a word. That word had been evoked in a skull box which, with all its contained powers and lurking paradoxes, has arisen in ways we can only dimly retrace.

Einstein is reputed to have once remarked that he refused to believe that God plays at dice with the universe. But as we survey the long backward course of history, it would appear that in the phenomenal world the open-endedness of time is unexpectedly an essential element of His creation. Whenever an infant is born, the dice, in the shape of genes and enzymes and the intangibles of chance environment, are being rolled again, as when that smoky figure from the fire hissed in my ear the tragedy of the cast-off infants of the city. Each one of us is a statistical impossibility around which hover a million other lives that were never destined to be born—but who, nevertheless, are being unmanifest, a lurking potential in the dark storehouse of the void.

Today, in spite of that web of law, that network of forces which the past century sought to string to the ends of the universe, a strange unexpectedness lingers about our world. This change in viewpoint, which has frequently escaped our attention, can be illustrated in the remark of Heinrich Hertz, the nineteenth-century experimenter in the electromagnetic field. "The most important problem which our conscious knowledge of nature should enable us to solve," Hertz stated, "is the anticipation of future events, so that we may arrange our present affairs in accordance with such anticipation."

There is an attraction about this philosophy that causes it to linger in the lay mind and, as a short-term prospect, in the minds of many scientists and technologists. It implies a tidiness that is infinitely attractive to man, increasingly a homeless orphan lost in the vast abysses of space and time. Hertz's remark seems to offer surcease from uncertainty, power contained, the universe understood, the future apprehended before its emergence. The previous Elizabethan age, by contrast, had often attached to its legal documents a humble obeisance to life's uncertainties expressed in the phrase "by the mutability of fortune and favor." The men of Shakespeare's century may have known less of science, but they knew only too well what unexpected overthrow was implied in the frown of a monarch or a breath of the plague.

The twentieth century, on the other hand, surveys a totally new universe. That our cosmological conceptions bear a relationship to the past is obvious, that some of the power of which Hertz dreamed lies in our hands is all too evident, but never before in human history has the mind soared higher and seen less to cheer its complacency. We have heard much of science as the endless frontier, but we whose immediate ancestors were seekers of gold among great mountains and gloomy forests are easily susceptible to a simplistic conception of the word *frontier* as something conquerable in its totality. We assume that, with enough time and expenditure of energy, the ore will be extracted and the forests computed in board feet of lumber. A tamed wilderness will subject itself to man.

Not so the wilderness beyond the stars or concealed in the infinitesimal world beneath the atom. Wise reflection will lead us to recognize that we have come upon a different and less conquerable region. Forays across its border already suggest that man's dream of mastering all aspects of nature takes no account of his limitations in time and space or of his own senses, augmented though they may be by his technological devices. Even the thought that he can bring to bear upon that frontier is limited in quantity by the number of trained minds that can sustain such an adventure. Ever more expensive grow the tools with which research can be sustained, ever more diverse the social problems which that research, in its technological phase, promotes. To take one single example: who would have dreamed that a tube connecting two lenses of glass would pierce into the swarming depths of our being, force upon us incredible feats of sanitary engineering, master the plague, and create that giant upsurge out of unloosened nature that we call the population explosion?

The Roman Empire is a past event in history, yet by analogy it presents us with a small scale model comparable to the endless frontier of science. A great political and military machine had expanded outward to the limits of the known world. Its lines of communication grew ever more tenuous, taxes rose fantastically, the disaffected and alienated within its borders steadily increased. By the time of the barbarian invasions the vast structure was already dying of inanition. Yet that empire lasted far longer than the world of science has yet endured.

But what of the empire of science? Does not its word leap fast as light, is it not a creator of incalculable wealth, is not space its plaything? Its weapons are monstrous; its eye is capable of peering beyond millions of light-years. There is one dubious answer to this buoyant optimism: science is human; it is of human devising and manufacture. It has not prevented war; it has perfected it. It has not abolished cruelty or corruption; it has enabled these abominations to be practiced on a scale unknown before in human history.

Science is a solver of problems, but it is dealing with the limitless, just as, in a cruder way, were the Romans. Solutions to problems create problems; their solutions, in turn, multiply into additional problems that escape out of scientific hands like noxious insects into the interstices of the social fabric. The rate of growth is geometric, and the vibrations set up can even now be detected in our institutions. This is what the Scottish biologist D'Arcy Thompson called the evolution of contingency. It is no longer represented by the long, slow turn of world time as the geologist has known it. Contingency has escaped into human hands and flickers unseen behind every whirl of our machines, every pronouncement of political policy.

Each one of us before his death looks back upon a childhood whose ways now seem as remote as those of Rome. "Daddy," the small daughter of a friend of mine recently asked, "tell me how it was in olden days." As my kindly friend groped amidst his classical history, he suddenly realized with a slight shock that his daughter wanted nothing more than an account of his own childhood. It was forty years away and it was already "olden days." "There was a time," he said slowly to the enchanted child, "called the years of the Great Depression. In that time there was a very great deal to eat, but men could not buy it. Little girls were scarcer than now. You see," he said painfully, "their fathers could not afford them, and they were not born." He made a half-apologetic gesture to the empty room, as if to a gathering of small reproachful ghosts. "There was a monster we never understood called Overproduction. There were," and his voice trailed hopelessly into silence, "so many dragons in that time you could not believe it. And there was a very civilized nation where little girls were taken from their parents. . . ." He could not go on. The eyes from Auschwitz, he told me later, would not permit him.

5

Recently, I passed a cemetery in a particularly bleak countryside. Adjoining the multitude of stark upthrust gray stones was an incongruous row of six transparent telephone booths erected in that spot for reasons best known to the communications industry. Were they placed there for the midnight convenience of the dead, or for the midday visitors who might attempt speech with the silent people beyond the fence? It was difficult to determine, but I thought the episode suggestive of our dilemma.

An instrument for communication, erected by a powerful unseen intelligence, was at my command, but I suspect—although I was oddly averse to trying to find out—that the wires did not run in the proper direction, and that there was something disconnected or disjointed about the whole endeavor. It was, I fear, symbolic of an unexpected aspect of our universe, a universe that, however strung with connecting threads, is endowed with an open-ended and perverse quality we shall never completely master. Nature contains that which does not concern us, and has no intention of taking us into its confidence. It may provide us with receiving boxes of white bone as cunning in their way as the wired booths in the cemetery, but, like these, they appear to lack some essential ingredient of genuine connection. As we consider what appears to be the chance emergence of photosynthesis, which turns the light of a far star into green leaves, or the creation of the phenomenon of sex that causes the cards at the gaming table of life to be shuffled with increasing frequency and into ever more diverse combinations, it should be plain that nature contains the roiling unrest of a tornado. It is not the self-contained stately palace of the eighteenth-century philosophers, a palace whose doorstep was always in precisely the same position.

From the oscillating universe, beating like a gigantic heart, to the puzzling existence of antimatter, order, in a human sense, is at least partially an illusion. Ours, in reality, is the order of a time, and of an insignificant fraction of the cosmos, seen by the limited senses of a finite creature. Behind the appearances, as even one group of primitive philosophers, the Hopi, have grasped, lurks being unmanifest, whose range and number exceeds the real. This is why the unexpected will always confront us; this is why the endless frontier is really endless. This is why the half-formed chaos of the marsh moved me as profoundly as though a new prophetic shape induced by us had risen monstrously from dangling wire and crumpled cardboard.

We are more dangerous than we seem and more potent in our ability to materialize the unexpected that is drawn from our own minds. "Force maketh Nature more violent in the Returne," Francis Bacon had once written. In the end, this is her primary quality. Her creature man partakes of that essence, and it is well that he consider it in contemplation and not always in action. To the unexpected nature of the universe man owes his being. More than any other living creature he contains, unknowingly, the shapes and forms of an uncreated future to be drawn from his own substance. The history of this unhappy century should prove a drastic warning of his powers of dissolution, even when directed upon himself. Waste, uncertain marshes, lie close to reality in our heads. Shapes as yet unevoked had best be left lying amidst those spectral bog lights, lest the drifting smoke of dreams merge imperceptibly, as once it did, with the choking real fumes from the ovens of Belsen and Buchenwald.

"It is very unhappy, but too late to be helped," Emerson had noted in his journal, "the discovery we have made that we exist. That discovery is called the Fall of Man. Ever afterwards we suspect our instruments. We have learned that we do not see directly." Wisdom interfused with compassion should be the consequence of that discovery, for at the same moment one aspect of the unexpected universe will have been genuinely revealed. It lies deep-hidden in the human heart, and not at the peripheries of space. Both the light we seek and the shadows that we fear are projected from within. It is through ourselves that the organic procession pauses, hesitates, or renews its journey. "We have learned to ask terrible questions," exclaimed that same thinker in the dawn of Victorian science. Perhaps it is just for this that the Unseen Player in the void has rolled his equally terrible dice. Out of the self-knowledge gained by putting dreadful questions man achieves his final dignity.

THREE

The Hidden Teacher

Sometimes the best teacher teaches only once to a single child or to a grownup past hope.

—ANONYMOUS

THE putting of formidable riddles did not arise with today's philosophers. In fact, there is a sense in which the experimental method of science might be said merely to have widened the area of man's homelessness. Over two thousand years ago, a man named Job, crouching in the Judean desert, was moved to challenge what he felt to be the injustice of his God. The voice in the whirlwind, in turn, volleyed pitiless questions upon the supplicant—questions that have, in truth, precisely the ring of modern science. For the Lord asked of Job by whose wisdom the hawk soars, and who had fathered the rain, or entered the storehouses of the snow.

A youth standing by, one Elihu, also played a role

in this drama, for he ventured diffidently to his protesting elder that it was not true that God failed to manifest Himself. He may speak in one way or another, though men do not perceive it. In consequence of this remark perhaps it would be well, whatever our individual beliefs, to consider what may be called the hidden teacher, lest we become too much concerned with the formalities of only one aspect of the education by which we learn.

We think we learn from teachers, and we sometimes do. But the teachers are not always to be found in school or in great laboratories. Sometimes what we learn depends upon our own powers of insight. Moreover, our teachers may be hidden, even the greatest teacher. And it was the young man Elihu who observed that if the old are not always wise, neither can the teacher's way be ordered by the young whom he would teach.

For example, I once received an unexpected lesson from a spider.

It happened far away on a rainy morning in the West. I had come up a long gulch looking for fossils, and there, just at eye level, lurked a huge yellow-and-black orb spider, whose web was moored to the tall spears of buffalo grass at the edge of the arroyo. It was her universe, and her senses did not extend beyond the lines and spokes of the great wheel she inhabited. Her extended claws could feel every vibration throughout that delicate structure. She knew the tug of wind, the fall of a raindrop, the flutter of a trapped moth's wing. Down one spoke of the web ran a stout ribbon of gossamer on which she could hurry out to investigate her prey.

Curious, I took a pencil from my pocket and touched a strand of the web. Immediately there was a response. The web, plucked by its menacing occupant, began to vibrate until it was a blur. Anything that had brushed claw or wing against that amazing snare would be thoroughly entrapped. As the vibrations slowed, I could see the owner fingering her guidelines for signs of struggle. A pencil point was an intrusion into this universe for which no precedent existed. Spider was circumscribed by spider ideas; its universe was spider universe. All outside was irrational, extraneous, at best, raw material for spider. As I proceeded on my way along the gully, like a vast impossible shadow, I realized that in the world of spider I did not exist.

Moreover, I considered, as I tramped along, that to the phagocytes, the white blood cells, clambering even now with some kind of elementary intelligence amid the thin pipes and tubing of my body—creatures without whose ministrations I could not exist—the conscious "I" of which I was aware had no significance to these amoeboid beings. I was, instead, a kind of chemical web that brought meaningful messages to them, a natural environment seemingly immortal if they could have thought about it, since generations of them had lived and perished, and would continue to so live and die, in that odd fabric which contained my intelligence—a misty light that was beginning to seem floating and tenuous even to me.

I began to see that among the many universes in which the world of living creatures existed, some were large, some small, but that all, including man's, were in some way limited or finite. We were creatures of many different dimensions passing through each other's lives like ghosts through doors.

In the years since, my mind has many times returned to that far moment of my encounter with the orb spider. A message has arisen only now from the misty shreds of that webbed universe. What was it that had so troubled me about the incident? Was it that spidery indifference to the human triumph?

If so, that triumph was very real and could not be denied. I saw, had many times seen, both mentally and in the seams of exposed strata, the long backward stretch of time whose recovery is one of the great feats of modern science. I saw the drifting cells of the early seas from which all life, including our own, has arisen. The salt of those ancient seas is in our blood, its lime is in our bones. Every time we walk along a beach some ancient urge disturbs us so that we find ourselves shedding shoes and garments, or scavenging among seaweed and whitened timbers like the homesick refugees of a long war.

And war it has been indeed—the long war of life against its inhospitable environment, a war that has lasted for perhaps three billion years. It began with strange chemicals seething under a sky lacking in oxygen; it was waged through long ages until the first green plants learned to harness the light of the nearest star, our sun. The human brain, so frail, so perishable, so full of inexhaustible dreams and hungers, burns by the power of the leaf.

The hurrying blood cells charged with oxygen carry more of that element to the human brain than to any other part of the body. A few moments' loss of vital air and the phenomenon we know as consciousness goes down into the black night of inorganic things. The human body is a magical vessel, but its life is linked with an element it cannot produce. Only the green plant knows the secret of transforming the light that comes to us across the far reaches of space. There is no better illustration of the intricacy of man's relationship with other living things.

The student of fossil life would be forced to tell us that if we take the past into consideration the vast majority of earth's creatures—perhaps over ninety per cent—have vanished. Forms that flourished for a far longer time than man has existed upon earth have become either extinct or so transformed that their descendants are scarcely recognizable. The specialized perish with the environment that created them, the tooth of the tiger fails at last, the lances of men strike down the last mammoth.

In three billion years of slow change and groping effort only one living creature has succeeded in escaping the trap of specialization that has led in time to so much death and wasted endeavor. It is man, but the word should be uttered softly, for his story is not yet done.

With the rise of the human brain, with the appearance of a creature whose upright body enabled two limbs to be freed for the exploration and manipulation of his environment, there had at last emerged a creature with a specialization—the brain—that, paradoxically, offered escape from specialization. Many animals driven into the nooks and crannies of nature have achieved momentary survival only at the cost of later extinction.

Was it this that troubled me and brought my mind back to a tiny universe among the grass-blades, a spider's universe concerned with spider thought?

Perhaps.

The mind that once visualized animals on a cave wall is now engaged in a vast ramification of itself through time and space. Man has broken through the boundaries that control all other life. I saw, at last, the reason for my recollection of that great spider on the arroyo's rim, fingering its universe against the sky.

The spider was a symbol of man in miniature. The wheel of the web brought the analogy home clearly. Man, too, lies at the heart of a web, a web extending through the starry reaches of sidereal space, as well as backward into the dark realm of prehistory. His great eye upon Mount Palomar looks into a distance of millions of light-years, his radio ear hears the whisper of even more remote galaxies, he peers through the electron microscope upon the minute particles of his own being. It is a web no creature of earth has ever spun before. Like the orb spider, man lies at the heart of it, listening. Knowledge has given him the memory of earth's history beyond the time of his emergence. Like the spider's claw, a part of him touches a world he will never enter in the flesh. Even now, one can see him reaching forward into time with new machines, computing, analyzing, until elements of the shadowy future will also compose part of the invisible web he fingers.

Yet still my spider lingers in memory against the sunset sky. Spider thoughts in a spider universe—sensitive to raindrop and moth flutter, nothing beyond, nothing allowed for the unexpected, the inserted pencil from the world outside.

Is man at heart any different from the spider, I wonder: man thoughts, as limited as spider thoughts, contemplating now the nearest star with the threat of bringing with him the fungus rot from earth, wars, violence, the burden of a population he refuses to control, cherishing again his dream of the Adamic Eden he had pursued and lost in the green forests of America. Now it beckons again like a mirage from beyond the moon. Let man spin his web, I thought further; it is his nature. But I considered also the work of the phagocytes swarming in the rivers of my body, the unresting cells in their mortal universe. What is it we are a part of that we do not see, as the spider was not gifted to discern my face, or my little probe into her world?

We are too content with our sensory extensions, with the fulfillment of that ice age mind that began its journey amidst the cold of vast tundras and that pauses only briefly before its leap into space. It is no longer enough to see as a man sees—even to the ends of the universe. It is not enough to hold nuclear energy in one's hand like a spear, as a man would hold it, or to see the lightning, or times past, or time to come, as a man would see it. If we continue to do this, the great brain—the human brain—will be only a new version of the old trap, and nature is full of traps for the beast that cannot learn.

It is not sufficient any longer to listen at the end of a wire to the rustlings of galaxies; it is not enough even to examine the great coil of DNA in which is coded the very alphabet of life. These are our extended perceptions. But beyond lies the great darkness of the ultimate Dreamer, who dreamed the light and the galaxies. Before act was, or substance existed, imagination grew in the dark. Man partakes of that ultimate wonder and creativeness. As we turn from the galaxies to the swarming cells of our own being, which toil for something, some entity beyond their grasp, let us remember man, the self-fabricator who came across an ice age to look into the mirrors and the magic of science. Surely he did not come to see himself or his wild visage only. He came because he is at heart a listener and a searcher for some transcendent realm beyond himself. This he has worshiped by many names, even in the dismal caves of his beginning. Man, the self-fabricator, is so by reason of gifts he had no part in devising—and so he searches as the single living cell in the beginning must have sought the ghostly creature it was to serve.

2

The young man Elihu, Job's counselor and critic, spoke simply of the "Teacher," and it is of this teacher I speak when I refer to gifts man had no part in devising. Perhaps—though it is purely a matter of emotional reactions to words—it is easier for us today to speak of this teacher as "nature," that omnipresent all which contained both the spider and my invisible intrusion into her carefully planned universe. But nature does not simply represent reality. In the shapes of life, it prepares the future; it offers alternatives. Nature teaches, though what it teaches is often hidden and obscure, just as the voice from the spinning dust cloud belittled Job's thought but gave back no answers to its own formidable interrogation.

A few months ago I encountered an amazing little creature on a windy corner of my local shopping center. It seemed, at first glance, some long-limbed, feathery spider teetering rapidly down the edge of a store front. Then it swung into the air and, as hesitantly as a spider on a thread, blew away into the parking lot. It returned in a moment on a gust of wind and ran toward me once more on its spindly legs with amazing rapidity.

With great difficulty I discovered the creature was actually a filamentous seed, seeking a hiding place and scurrying about with the uncanny surety of a conscious animal. In fact, it *did* escape me before I could secure it. Its flexible limbs were stiffer than milkweed down, and, propelled by the wind, it ran rapidly and evasively over the pavement. It was like a gnome scampering somewhere with a hidden packet—for all that I could tell, a totally new one: one of the jumbled alphabets of life.

A new one? So stable seem the years and all green leaves, a botanist might smile at my imaginings. Yet bear with me a moment. I would like to tell a tale, a genuine tale of childhood. Moreover, I was just old enough to know the average of my kind and to marvel at what I saw. And what I saw was straight from the hidden Teacher, whatever be his name.

It is told in the Orient of the Hindu god Krishna that his mother, wiping his mouth when he was a child, inadvertently peered in and beheld the universe, though the sight was mercifully and immediately veiled from her. In a sense, this is what happened to me. One day there arrived at our school a newcomer, who entered the grade above me. After some days this lad, whose look of sleepy-eyed arrogance is still before me as I write, was led into my mathematics classroom by the principal. Our class was informed severely that we should learn to work harder.

With this preliminary exhortation, great rows of figures were chalked upon the blackboard, such difficult mathematical problems as could be devised by adults. The class watched in helpless wonder. When the preparations had been completed, the young pupil sauntered forward and, with a glance of infinite boredom that swept from us to his fawning teachers, wrote the answers, as instantaneously as a modern computer, in their proper place upon the board. Then he strolled out with a carelessly exaggerated yawn.

Like some heavy-browed child at the wood's edge, clutching the last stone hand ax, I was witnessing the birth of a new type of humanity—one so beyond its teachers that it was being used for mean purposes while the intangible web of the universe in all its shimmering mathematical perfection glistened untaught in the mind of a chance little boy. The boy, by then grown self-centered and contemptuous, was being dragged from room to room to encourage us, the paleanthropes, to duplicate what, in reality, our teachers could not duplicate. He was too precious an object to be released upon the playground among us, and with reason. In a few months his parents took him away.

Long after, looking back from maturity, I realized that I had been exposed on that occasion, not to human teaching, but to the Teacher, toying with some sixteen billion nerve cells interlocked in ways past understanding. Or, if we do not like the anthropomorphism implied in the word *teacher,* then nature, the old voice from the whirlwind fumbling for the light. At all events, I had been the fortunate witness to life's unbounded creativity—a creativity seemingly still as unbalanced and chance-filled as in that far era when a black scaled creature had broken from an egg and the age of the giant reptiles, the creatures of the prime, had tentatively begun.

Because form cannot be long sustained in the living, we collapse inward with age. We die. Our bodies, which were the product of a kind of hidden teaching by an alphabet we are only beginning dimly to discern, are dismissed into their elements. What is carried onward, assuming we have descendants, is the little capsule of instructions such as I encountered hastening by me in the shape of a running seed. We have learned the first biological lesson: that in each generation life passes through the eye of a needle. It exists for a time molecularly and in no recognizable semblance to its adult condition. It *instructs* its way again into man or reptile. As the ages pass, so do variants of the code. Occasionally, a species vanishes on a wind as unreturning as that which took the pterodactyls.

Or the code changes by subtle degrees through the statistical altering of individuals; until I, as the fading Neanderthals must once have done, have looked with still-living eyes upon the creature whose genotype was quite possibly to replace me. The genetic alphabets, like genuine languages, ramify and evolve along unreturning pathways.

If nature's instructions are carried through the eye of a needle, through the molecular darkness of a minute world below the field of human vision and of time's decay, the same, it might be said, is true of those monumental structures known as civilizations. They are transmitted from one generation to another on invisible puffs of air known as words—words that can also be symbolically incised on clay. As the delicate printing on the mud at the water's edge retraces a visit of autumn birds long since departed, so the little scrabbled tablets in perished cities carry the seeds of human thought across the deserts of millennia. In this instance the teacher is the social brain but it, too, must be compressed into minute hieroglyphs, and the minds that wrought the miracle efface themselves amidst the jostling torrent of messages, which, like the genetic code, are shuffled and reshuffled as they hurry through eternity. Like a mutation, an idea may be recorded in the wrong time, to lie latent like a recessive gene and spring once more to life in an auspicious era.

Occasionally, in the moments when an archaeologist lifts the slab over a tomb that houses a great secret, a few men gain a unique glimpse through that dark portal out of which all men living have emerged, and through which messages again must pass. Here the Mexican archaeologist Ruz Lhuillier speaks of his first penetration of the great tomb hidden beneath dripping stalactites at the pyramid of Palenque: "Out of the dark shadows, rose a fairy-tale vision, a weird ethereal spectacle from another world. It was like a magician's cave carved out of ice, with walls glittering and sparkling like snow crystals." After shining his torch over hieroglyphs and sculptured figures, the explorer remarked wonderingly: "We were the first people for more than a thousand years to look at it."

Or again, one may read the tale of an unknown Pharaoh who had secretly arranged that a beloved woman of his household should be buried in the tomb of the god-king—an act of compassion carrying a personal message across the millennia in defiance of all precedent.

Up to this point we have been talking of the single hidden teacher, the taunting voice out of that old

Biblical whirlwind which symbolizes nature. We have seen incredible organic remembrance passed through the needle's eye of a microcosmic world hidden completely beneath the observational powers of creatures preoccupied and ensorcelled by dissolution and decay. We have seen the human mind unconsciously seize upon the principles of that very code to pass its own societal memory forward into time. The individual, the momentary living cell of the society, vanishes, but the institutional structures stand, or if they change, do so in an invisible flux not too dissimilar from that persisting in the stream of genetic continuity.

Upon this world, life is still young, not truly old as stars are measured. Therefore it comes about that we minimize the role of the synapsid reptiles, our remote forerunners, and correspondingly exalt our own intellectual achievements. We refuse to consider that in the old eye of the hurricane we may be, and doubtless are, in aggregate, a slightly more diffuse and dangerous dragon of the primal morning that still enfolds us.

Note that I say "in aggregate." For it is just here, among men, that the role of messages, and, therefore, the role of the individual teacher—or, I should say now, the hidden teachers—begin to be more plainly apparent and their instructions become more diverse. The dead Pharaoh, though unintentionally, by a revealing act, had succeeded in conveying an impression of human tenderness that has outlasted the trappings of a vanished religion.

Like most modern educators I have listened to student demands to grade their teachers. I have heard the words repeated until they have become a slogan, that no man over thirty can teach the young of this generation. How would one grade a dead Pharaoh, millennia gone, I wonder, one who did not intend to teach, but who, to a few perceptive minds, succeeded by the simple nobility of an act.

Many years ago, a student who was destined to become an internationally known anthropologist sat in a course in linguistics and heard his instructor, a man of no inconsiderable wisdom, describe some linguistic peculiarities of Hebrew words. At the time, the young student, at the urging of his family, was contemplating a career in theology. As the teacher warmed to his subject, the student, in the back row, ventured excitedly:

"I believe I can understand that, sir. It is very similar to what exists in Mohegan."

The linguist paused and adjusted his glasses. "Young man," he said, "Mohegan is a dead language. Nothing has been recorded of it since the eighteenth century. Don't bluff."

"But sir," the young student countered hopefully, "it can't be dead so long as an old woman I know still speaks it. She is Pequot-Mohegan. I learned a bit of vocabulary from her and could speak with her myself. She took care of me when I was a child."

"Young man," said the austere, old-fashioned scholar, "be at my house for dinner at six this evening. You and I are going to look into this matter."

A few months later, under careful guidance, the young student published a paper upon Mohegan linguistics, the first of a long series of studies upon the forgotten languages and ethnology of the Indians of the Northeastern forests. He had changed his vocation and turned to anthropology because of the attraction of a hidden teacher. But just who was the teacher? The young man himself, his instructor, or that solitary speaker of a dying tongue who had so yearned to hear her people's voice that she had softly babbled it to a child?

Later, this man was to become one of my professors. I absorbed much from him, though I hasten to make the reluctant confession that he was considerably beyond thirty. Most of what I learned was gathered over cups of coffee in a dingy campus restaurant. What we talked about were things some centuries older than either of us. Our common interest lay in snakes, scapulimancy, and other forgotten rites of benighted forest hunters.

I have always regarded this man as an extraordinary individual, in fact, a hidden teacher. But alas, it is all now so old-fashioned. We never protested the impracticality of his quaint subjects. We were all too ready to participate in them. He was an excellent canoeman, but he took me to places where I fully expected to drown before securing my degree. To this day, fragments of his unused wisdom remain stuffed in some back attic of my mind. Much of it I have never found the opportunity to employ, yet it has somehow colored my whole adult existence. I belong to that elderly professor in somewhat the same way that he, in turn, had become the wood child of a hidden forest mother.

There are, however, other teachers. For example, among the hunting peoples there were the animal counselors who appeared in prophetic dreams. Or, among the Greeks, the daemonic supernaturals who stood at the headboard while a man lay stark and listened—sometimes to dreadful things. "You are asleep," the messengers proclaimed over and over again, as though the man lay in a spell to hear his doom pronounced. "You, Achilles, you, son of Atreus. You are asleep, asleep," the hidden ones pronounced and vanished.

We of this modern time know other things of dreams, but we know also that they can be interior teachers and healers as well as the anticipators of disaster. It has been said that great art is the night thought of man. It may emerge without warning from the soundless depths of the unconscious, just as supernovae may blaze up suddenly in the farther reaches of void space. The critics, like astronomers, can afterward triangulate such worlds but not account for them.

A writer friend of mine with bitter memories of his youth and estranged from his family, who, in the interim, had died, gave me this account of the matter in his middle years. He had been working, with an unusual degree of reluctance, upon a novel that contained certain autobiographical episodes. One night he dreamed; it was a very vivid and stunning dream in its detailed reality.

He found himself hurrying over creaking snow through the blackness of a winter night. He was ascending a familiar path through a long-vanished orchard.

The path led to his childhood home. The house, as he drew near, appeared dark and uninhabited, but, impelled by the power of the dream, he stepped upon the porch and tried to peer through a dark window into his own old room.

"Suddenly," he told me, "I was drawn by a strange mixture of repulsion and desire to press my face against the glass. I knew intuitively they were all there waiting for me within, if I could but see them. My mother and my father. Those I had loved and those I hated. But the window was black to my gaze. I hesitated a moment and struck a match. For an instant in that freezing silence I saw my father's face glimmer wan and remote behind the glass. My mother's face was there, with the hard, distorted lines that marked her later years.

"A surge of fury overcame my cowardice. I cupped the match before me and stepped closer, closer toward that dreadful confrontation. As the match guttered down, my face was pressed almost to the glass. In some quick transformation, such as only a dream can effect, I saw that it was my own face into which I stared, just as it was reflected in the black glass. My father's haunted face was but my own. The hard lines upon my mother's aging countenance were slowly reshaping themselves upon my living face. The light burned out. I awoke sweating from the terrible psychological tension of that nightmare. I was in a far port in a distant land. It was dawn. I could hear the waves breaking on the reef."

"And how do you interpret the dream?" I asked,

concealing a sympathetic shudder and sinking deeper into my chair.

"It taught me something," he said slowly, and with equal slowness a kind of beautiful transfiguration passed over his features. All the tired lines I had known so well seemed faintly to be subsiding.

"Did you ever dream it again?" I asked out of a comparable experience of my own.

"No, never," he said, and hesitated. "You see, I had learned it was just I, but more, much more, I had learned that I was they. It makes a difference. And at the last, late, much too late, it was all right. I understood. My line was dying, but I understood. I hope they understood, too." His voice trailed into silence.

"It is a thing to learn," I said. "You were seeking something and it came." He nodded, wordless. "Out of a tomb," he added after a silent moment, "my kind of tomb—the mind."

On the dark street, walking homeward, I considered my friend's experience. Man, I concluded, may have come to the end of that wild being who had mastered the fire and the lightning. He can create the web but not hold it together, not save himself except by transcending his own image. For at the last, before the ultimate mystery, it is himself he shapes. Perhaps it is for this that the listening web lies open: that by knowledge we may grow beyond our past, our follies, and ever closer to what the Dreamer in the dark intended before the dust arose and walked. In the pages of an old book it has been written that we are in the hands of a Teacher, nor does it yet appear what man shall be.

FOUR

The Star Thrower

Who is the man walking in the Way?
An eye glaring in the skull.

—SECCHO

IT has ever been my lot, though formally myself a teacher, to be taught surely by none. There are times when I have thought to read lessons in the sky, or in books, or from the behavior of my fellows, but in the end my perceptions have frequently been inadequate or betrayed. Nevertheless, I venture to say that of what man may be I have caught a fugitive glimpse, not among multitudes of men, but along an endless wave-beaten coast at dawn. As always, there is this apparent break, this rift in nature, before the insight comes. The terrible question has to translate itself into an even more terrifying freedom.

If there is any meaning to this book, it began on the beaches of Costabel with just such a leap across an unknown abyss. It began, if I may borrow the expression from a Buddhist sage, with the skull and the eye. I was the skull. I was the inhumanly stripped skeleton without voice, without hope, wandering alone upon the shores of the world. I was devoid of pity, because pity implies hope. There was, in this desiccated skull, only an eye like a pharos light, a beacon, a search beam revolving endlessly in sunless noonday or black night. Ideas like swarms of insects rose to the beam, but the light consumed them. Upon that shore meaning had ceased. There were only the dead skull and the revolving eye. With such an eye, some have said, science looks upon the world. I do not know. I know only that I was the skull of emptiness and the endlessly revolving light without pity.

Once, in a dingy restaurant in the town, I had heard a woman say: "My father reads a goose bone for the weather." A modern primitive, I had thought, a diviner, using a method older than Stonehenge, as old as the arctic forests.

"And where does he do that?" the woman's companion had asked amusedly.

"In Costabel," she answered complacently, "in Costabel." The voice came back and buzzed faintly for a moment in the dark under the revolving eye. It did not make sense, but nothing in Costabel made sense. Perhaps that was why I had finally found myself in Costabel. Perhaps all men are destined at some time to arrive there as I did.

I had come by quite ordinary means, but I was still the skull with the eye. I concealed myself beneath a fisherman's cap and sunglasses, so that I looked like everyone else on the beach. This is the way things are managed in Costabel. It is on the shore that the revolving eye begins its beam and the whispers rise in the empty darkness of the skull.

The beaches of Costabel are littered with the debris of life. Shells are cast up in windrows; a hermit crab, fumbling for a new home in the depths, is tossed naked ashore, where the waiting gulls cut him to pieces. Along the strip of wet sand that marks the ebbing and flowing of the tide death walks hugely and in many forms. Even the torn fragments of green sponge yield bits of scrambling life striving to return to the great mother that has nourished and protected them.

In the end the sea rejects its offspring. They cannot fight their way home through the surf which casts them repeatedly back upon the shore. The tiny breathing pores of starfish are stuffed with sand. The rising sun shrivels the mucilaginous bodies of the unprotected. The seabeach and its endless war are soundless. Nothing screams but the gulls.

In the night, particularly in the tourist season, or during great storms, one can observe another vulturine activity. One can see, in the hour before dawn on the ebb tide, electric torches bobbing like fireflies along the beach. It is the sign of the professional shellers seeking to outrun and anticipate their less aggressive neighbors. A kind of greedy madness sweeps over the competing collectors. After a storm one can see them hurrying along with bundles of gathered starfish, or, toppling and overburdened, clutching bags of living shells whose hidden occupants will be slowly cooked and dissolved in the outdoor kettles provided by the resort hotels for the cleaning of specimens. Following one such episode I met the star thrower.

As soon as the ebb was flowing, as soon as I could make out in my sleeplessness the flashlights on the beach, I arose and dressed in the dark. As I came down the steps to the shore I could hear the deeper rumble of the surf. A gaping hole filled with churning sand had cut sharply into the breakwater. Flying sand as light as powder coated every exposed object like snow. I made my way around the altered edges of the cove and proceeded on my morning walk up the shore. Now and then a stooping figure moved in the gloom or a rain squall swept past me with light pattering steps. There was a faint sense of coming light somewhere behind me in the east.

Soon I began to make out objects, upended timbers, conch shells, sea wrack wrenched from the far-out kelp forests. A pink-clawed crab encased in a green cup of sponge lay sprawling where the waves had tossed him. Long-limbed starfish were strewn everywhere, as though the night sky had showered down. I paused once briefly. A small octopus, its beautiful dark-lensed eyes bleared with sand, gazed up at me from a ragged bundle of tentacles. I hesitated, and touched it briefly with my foot. It was dead. I paced on once more before the spreading whitecaps of the surf.

The shore grew steeper, the sound of the sea heavier and more menacing, as I rounded a bluff into the full blast of the offshore wind. I was away from the shellers now and strode more rapidly over the wet sand that effaced my footprints. Around the next point there might be a refuge from the wind. The sun behind me was pressing upward at the horizon's rim—an ominous red glare amidst the tumbling blackness of the clouds. Ahead of me, over the projecting point, a gigantic rainbow of incredible perfection had sprung shimmering into existence. Somewhere toward its foot I discerned a human figure standing, as it seemed to me, within the rainbow, though unconscious of his position. He was gazing fixedly at something in the sand.

Eventually he stooped and flung the object beyond the breaking surf. I labored toward him over a half mile of uncertain footing. By the time I reached him the rainbow had receded ahead of us, but something of its color still ran hastily in many changing lights across his features. He was starting to kneel again.

In a pool of sand and silt a starfish had thrust its arms up stiffly and was holding its body away from the stifling mud.

"It's still alive," I ventured.

"Yes," he said, and with a quick yet gentle movement he picked up the star and spun it over my head and far out into the sea. It sank in a burst of spume, and the waters roared once more.

"It may live," he said, "if the offshore pull is strong enough." He spoke gently, and across his bronzed worn face the light still came and went in subtly altering colors.

"There are not many come this far," I said, groping in a sudden embarrassment for words. "Do you collect?"

"Only like this," he said softly, gesturing amidst the wreckage of the shore. "And only for the living." He stooped again, oblivious of my curiosity, and skipped another star neatly across the water.

"The stars," he said, "throw well. One can help them."

He looked full at me with a faint question kindling in his eyes, which seemed to take on the far depths of the sea.

"I do not collect," I said uncomfortably, the wind beating at my garments. "Neither the living nor the dead. I gave it up a long time ago. Death is the only successful collector." I could feel the full night blackness in my skull and the terrible eye resuming its indifferent journey. I nodded and walked away, leaving him there upon the dune with that great rainbow ranging up the sky behind him.

I turned as I neared a bend in the coast and saw him toss another star, skimming it skillfully far out over the ravening and tumultuous water. For a moment, in the changing light, the sower appeared magnified, as though casting larger stars upon some greater sea. He had, at any rate, the posture of a god.

But again the eye, the cold world-shriveling eye, began its inevitable circling in my skull. He is a man, I considered sharply, bringing my thought to rest. The star thrower is a man, and death is running more fleet than he along every seabeach in the world.

I adjusted the dark lens of my glasses and, thus disguised, I paced slowly back by the starfish gatherers, past the shell collectors, with their vulgar little spades and the stick-length shelling pincers that eased their elderly backs while they snatched at treasures in the sand. I chose to look full at the steaming kettles in which beautiful voiceless things were being boiled alive. Behind my sunglasses a kind of litany began and refused to die down. *"As I came through the desert thus it was, as I came through the desert."*

In the darkness of my room I lay quiet with the sunglasses removed, but the eye turned and turned. In the desert, an old monk had once advised a traveler, the voices of God and the Devil are scarcely distinguishable. Costabel was a desert. I lay quiet, but my restless hand at the bedside fingered the edge of an invisible abyss. "Certain coasts," the remark of a perceptive writer came back to me, "are set apart for shipwreck." With unerring persistence I had made my way thither.

2

There is a difference in our human outlook, depending on whether we have been born upon level plains, where one step reasonably leads to another, or whether, by contrast, we have spent our lives amidst glacial crevasses and precipitous descents. In the case of the mountaineer, one step does not always lead rationally to another save by a desperate leap over a chasm or by an even more hesitant tiptoeing across precarious snow bridges.

Something about these opposed landscapes has its analogue in the mind of man. Our prehistoric life, one might say, began amidst enforested gloom with the abandonment of the protected instinctive life of nature. We sought, instead, an adventurous existence amidst the crater lands and ice fields of self-generated ideas. Clambering onward, we have slowly made our way out of a maze of isolated peaks into the level plains of science. Here, one step seems definitely to succeed another, the universe appears to take on an imposed order, and the illusions through which mankind has painfully made its way for many centuries have given place to the enormous vistas of past and future time. The encrusted eye in the stone speaks to us of undeviating sunlight; the calculated elliptic of Halley's comet no longer forecasts world disaster. The planet plunges on through a chill void of star years, and there is little or nothing that remains unmeasured.

Nothing, that is, but the mind of man. Since boyhood I had been traveling across the endless co-ordinated realms of science, just as, in the body, I was a plains dweller, accustomed to plodding through distances unbroken by precipices. Now that I come to look back, there was one contingent aspect of that landscape I inhabited whose significance, at the time, escaped me. "Twisters," we called them locally. They were a species of cyclonic, bouncing air funnel that could suddenly loom out of nowhere, crumpling windmills or slashing with devastating fury through country towns. Sometimes, by modest contrast, more harmless varieties known as dust devils might pursue one in a gentle spinning dance for miles. One could see them hesitantly stalking across the alkali flats on a hot day, debating, perhaps, in their tall, rotating columns, whether to ascend and assume more formidable shapes. They were the trickster part of an otherwise pedestrian landscape.

Infrequent though the visitations of these malign creations of the air might be, all prudent homesteaders in those parts had provided themselves with cyclone cellars. In the careless neighborhood in which I grew up, however, we contented ourselves with the queer yarns of cyclonic folklore and the vagaries of weather prophecy. As a boy, aroused by these tales and cherishing a subterranean fondness for caves, I once attempted to dig a storm cellar. Like most such projects this one was never completed. The trickster element in nature, I realize now, had so buffeted my parents that they stoically rejected planning. Unconsciously, they had arrived at the philosophy that foresight merely invited the attention of some baleful intelligence that despised and persecuted the calculating planner. It was not until many years later that I came to realize that a kind of maleficent primordial power persists in the mind as well as in the wandering dust storms of the exterior world.

A hidden dualism that has haunted man since antiquity runs across his religious conceptions as the conflict between good and evil. It persists in the modern world of science under other guises. It becomes chaos versus form or antichaos. Form, since the rise of the evolutionary philosophy, has itself taken on an illusory quality. Our apparent shapes no longer have the stability of a single divine fiat. Instead, they waver and dissolve into the unexpected. We gaze backward into a contracting cone of life until words leave us and all we know is dissolved into the simple circuits of a reptilian brain. Finally, sentience subsides into an animalcule.

Or we revolt and refuse to look deeper, but the void remains. We are rag dolls made out of many ages and skins, changelings who have slept in wood nests or hissed in the uncouth guise of waddling amphibians. We have played such roles for infinitely longer ages than we have been men. Our identity is a dream. We are process, not reality, for reality is an illusion of the daylight—the light of our particular day. In a fortnight, as aeons are measured, we may lie silent in a bed of stone, or, as has happened in the past, be figured in another guise. Two forces struggle perpetually in our bodies: Yam, the old sea dragon of the original Biblical darkness, and, arrayed against him, some wisp of dancing light that would have us linger, wistful, in our human form. "Tarry thou, till I come again"—an old legend survives among us of the admonition given by Jesus to the Wandering Jew. The words are applicable to all of us. Deep-hidden in the human psyche there is a similar injunction no longer having to do with the longevity of the body but, rather, a plea to wait upon some transcendent lesson preparing in the mind itself.

Yet the facts we face seem terrifyingly arrayed

against us. It is as if at our backs, masked and demonic, moved the trickster as I have seen his role performed among the remnant of a savage people long ago. It was that of the jokester present at the most devout of ceremonies. This creature never laughed; he never made a sound. Painted in black, he followed silently behind the officiating priest, mimicking, with the added flourish of a little whip, the gestures of the devout one. His timed and stylized posturings conveyed a derision infinitely more formidable than actual laughter.

In modern terms, the dance of contingency, of the indeterminable, outwits us all. The approaching, fateful whirlwind on the plain had mercifully passed me by in youth. In the moment when I had witnessed that fireside performance I knew with surety that primitive man had lived with a dark message. He had acquiesced in the admission into his village of a cosmic messenger. Perhaps the primitives were wiser in the ways of the trickster universe than ourselves; perhaps they knew, as we do not, how to ground or make endurable the lightning.

At all events, I had learned, as I watched that half-understood drama by the leaping fire, why man, even modern man, reads goose bones for the weather of his soul. Afterward I had gone out, a troubled unbeliever, into the night. There was a shadow I could not henceforth shake off, which I knew was posturing and would always posture behind me. That mocking shadow looms over me as I write. It scrawls with a derisive pen and an exaggerated flourish. I know instinctively it will be present to caricature the solemnities of my deathbed. In a quarter of a century it has never spoken.

Black magic, the magic of the primeval chaos, blots out or transmogrifies the true form of things. At the stroke of twelve the princess must flee the banquet or risk discovery in the rags of a kitchen wench; coach reverts to pumpkin. Instability lies at the heart of the world. With uncanny foresight folklore has long toyed symbolically with what the nineteenth century was to proclaim a reality, namely, that form is an illusion of the time dimension, that the magic flight of the pursued hero or heroine through frogskin and wolf coat has been, and will continue to be, the flight of all men.

Goethe's genius sensed, well before the publication of the *Origin of Species,* the thesis and antithesis that epitomize the eternal struggle of the immediate species against its dissolution into something other: in modern terms, fish into reptile, ape into man. The power to change is both creative and destructive—a sinister gift, which, unrestricted, leads onward toward the formless and inchoate void of the possible. This force can only be counterbalanced by an equal impulse toward specificity. Form, once arisen, clings to its identity. Each species and each individual holds tenaciously to its present nature. Each strives to contain the creative and abolishing maelstrom that pours unseen through the generations. The past vanishes; the present momentarily persists; the future is potential only. In this specious present of the real, life struggles to maintain every manifestation, every individuality, that exists. In the end, life always fails, but the amorphous hurrying stream is held and diverted into new organic vessels in which form persists, though the form may not be that of yesterday.

The evolutionists, piercing beneath the show of momentary stability, discovered, hidden in rudimentary organs, the discarded rubbish of the past. They detected the reptile under the lifted feathers of the bird, the lost terrestrial limbs dwindling beneath the blubber of the giant cetaceans. They saw life rushing outward from an unknown center, just as today the astronomer senses the galaxies fleeing into the infinity of darkness. As the spinning galactic clouds hurl stars and worlds across the night, so life, equally impelled by the centrifugal powers lurking in the germ cell, scatters the splintered radiance of consciousness and sends it prowling and contending through the thickets of the world.

All this devious, tattered way was exposed to the ceaselessly turning eye within the skull that lay hidden upon the bed in Costabel. Slowly that eye grew conscious of another eye that searched it with equal penetration from the shadows of the room. It may have been a projection from the mind within the skull, but the eye was, nevertheless, exteriorized and haunting. It began as something glaucous and blind beneath a web of clinging algae. It altered suddenly and became the sand-smeared eye of the dead cephalopod I had encountered upon the beach. The transformations became more rapid with the concentration of my attention, and they became more formidable. There was the beaten, bloodshot eye of an animal from somewhere within my childhood experience. Finally, there was an eye that seemed torn from a photograph, but that looked through me as though it had already raced in vision up to the steep edge of nothingness and absorbed whatever terror lay in that abyss. I sank back again upon my cot and buried my head in the pillow. I knew the eye and the circumstance and the question. It was my mother. She was long dead, and the way backward was lost.

3

Now it may be asked, upon the coasts that invite shipwreck, why the ships should come, just as we may ask the man who pursues knowledge why he should be left with a revolving search beam in the head whose light falls only upon disaster or the flotsam of the shore. There is an answer, but its way is not across the level plains of science, for the science of remote abysses no longer shelters man. Instead, it reveals him in vaporous metamorphic succession as the homeless and unspecified one, the creature of the magic flight.

Long ago, when the future was just a simple tomorrow, men had cast intricately carved game counters to determine its course, or they had traced with a grimy finger the cracks on the burnt shoulder blade of a hare. It was a prophecy of tomorrow's hunt, just as was the old farmer's anachronistic reading of the weather from the signs on the breastbone of a goose. Such quaint almanacs of nature's intent had sufficed mankind since antiquity. They would do so no longer, nor would formal apologies to the souls of the game men hunted. The hunters had come, at last, beyond the satisfying supernatural world that had always surrounded the little village, into a place of homeless frontiers and precipitous edges, the indescribable world of the natural. Here tools increasingly revenged themselves upon their creators and tomorrow became unmanageable. Man had come in his journeying to a region of terrible freedoms.

It was a place of no traditional shelter, save those erected with the aid of tools, which had also begun to achieve a revolutionary independence from their masters. Their ways had grown secretive and incalculable. Science, more powerful than the magical questions that might be addressed by a shaman to a burnt shoulder blade, could create these tools but had not succeeded in controlling their ambivalent nature. Moreover, they responded all too readily to that urge for tampering and dissolution which is part of our primate heritage.

We had been safe in the enchanted forest only because of our weakness. When the powers of that gloomy region were given to us, immediately, as in a witch's house, things began to fly about unbidden. The tools, if not science itself, were linked intangibly to the subconscious poltergeist aspect of man's nature. The closer man and the natural world drew together, the more erratic became the behavior of each. Huge shadows leaped triumphantly after every blinding illumination.

It was a magnified but clearly recognizable version of the black trickster's antics behind the solemn backs of the priesthood. Here, there was one difference. The shadows had passed out of all human semblance; no societal ritual safely contained their posturings, as in the warning dance of the trickster. Instead, unseen by many because it was so gigantically real, the multiplied darkness threatened to submerge the carriers of the light.

Darwin, Einstein, and Freud might be said to have released the shadows. Yet man had already entered the perilous domain that henceforth would contain his destiny. Four hundred years ago Francis Bacon had already anticipated its dual nature. The individuals do not matter. If they had not made their discoveries, others would have surely done so. They were good men, and they came as enlighteners of mankind. The tragedy was only that at their backs the ritual figure with the whip was invisible. There was no longer anything to subdue the pride of man. The world had been laid under the heavy spell of the natural; henceforth, it would be ordered by man.

Humanity was suddenly entranced by light and fancied it reflected light. Progress was its watchword, and for a time the shadows seemed to recede. Only a few guessed that the retreat of darkness presaged the emergence of an entirely new and less tangible terror. Things, in the words of G. K. Chesterton, were to grow incalculable by being calculated. Man's powers were finite; the forces he had released in nature recognized no such limitations. They were the irrevocable monsters conjured up by a completely amateur sorcerer.

But what, we may ask, was the nature of the first discoveries that now threaten to induce disaster? Preeminent among them was, of course, the perception to which we have already referred: the discovery of the interlinked and evolving web of life. The great Victorian biologists saw, and yet refused to see, the war between form and formlessness, chaos and antichaos, which the poet Goethe had sensed contesting beneath the smiling surface of nature. "The dangerous gift from above," he had termed it, with uneasy foresight.

By contrast, Darwin, the prime student of the struggle for existence, sought to visualize in a tangled bank of leaves the silent and insatiable war of nature. Still, he could imply with a veiled complacency that man might "with some confidence" look forward to a secure future "of inappreciable length." This he could do upon the same page in the *Origin of Species* where he observes that "of the species now living very few will transmit progeny to a far distant futurity." The contradiction escaped him; he did not wish to see it. Darwin, in addition, saw life as a purely selfish struggle, in which nothing is modified for the good of another species without being directly advantageous to its associated form.

If, he contended, one part of any single species had been formed for the exclusive good of another, "it would annihilate my theory." Powerfully documented and enhanced though the statement has become, famine, war, and death are not the sole instruments biologists today would accept as the means toward that perfection of which Darwin spoke. The subject is subtle and intricate, however, and one facet of it must be reserved for another chapter. Let it suffice to say here that the sign of the dark cave and the club became so firmly fixed in human thinking that in our time it has been invoked as signifying man's true image in books selling in the hundreds of thousands.

From the thesis and antithesis contained in Darwinism we come to Freud. The public knows that, like Darwin, the master of the inner world took the secure, stable, and sunlit province of the mind and revealed it as a place of contending furies. Ghostly transformations, flitting night shadows, misshapen changelings existed there, as real as anything that haunted the natural universe of Darwin. For this reason, appropriately, I had come as the skull and the eye to Costabel—the coast demanding shipwreck. Why else had I remembered the phrase, except for a dark impulse toward destruction lurking somewhere in the subconscious? I lay on the bed while the agonized eye in the remembered photograph persisted at the back of my closed lids.

It had begun when, after years of separation, I had gone dutifully home to a house from which the final occupant had departed. In a musty attic—among old trunks, a broken aquarium, and a dusty heap of fossil shells collected in childhood—I found a satchel. The satchel was already a shabby antique, in whose depths I turned up a jackknife and a "rat" of hair such as women wore at the beginning of the century. Beneath these lay a pile of old photographs and a note—two notes, rather, evidently dropped into the bag at different times. Each, in a thin, ornate hand, reiterated a single message that the writer had believed important. "This satchel belongs to my son, Loren Eiseley." It was the last message. I recognized the trivia. The jackknife I had carried in childhood. The rat of hair had belonged to my mother, and there were also two incredibly pointed slippers that looked as though they had been intended for a formal ball, to which I knew well my mother would never in her life have been invited. I undid the rotted string around the studio portraits.

Mostly they consisted of stiff, upright bearded men and heavily clothed women equally bound to the formalities and ritual that attended upon the photography of an earlier generation. No names identified the pictures, although here and there a reminiscent family trait seemed faintly evident. Finally I came upon a less formal photograph, taken in the eighties of the last century. Again no names identified the people, but a commercial stamp upon the back identified the place: Dyersville, Iowa. I had never been in that country town, but I knew at once it was my mother's birthplace.

Dyersville, the thought flashed through my mind, making the connection now for the first time: the dire place. I recognized at once the two sisters at the edge of the photograph, the younger clinging reluctantly to the older. Six years old, I thought, turning momentarily away from the younger child's face. Here it began, her pain and mine. The eyes in the photograph were already remote and shadowed by some inner turmoil. The poise of the body was already that of one miserably departing the peripheries of the human estate. The gaze was mutely clairvoyant and lonely. It was the gaze of a child who knew unbearable difference and impending isolation.

I dropped the notes and pictures once more into the bag. The last message had come from Dyersville: "my son." The child in the photograph had survived to be an ill-taught prairie artist. She had been deaf. All her life she had walked the precipice of mental breakdown. Here on this faded porch it had begun—the long crucifixion of life. I slipped downstairs and out of the house. I walked for miles through the streets.

Now at Costabel I put on the sunglasses once more, but the face from the torn photograph persisted behind them. It was as though I, as man, was being asked to confront, in all its overbearing weight, the universe itself. "Love not the world," the Biblical injunction runs, "neither the things that are in the world." The revolving beam in my mind had stopped, and the insect whisperings of the intellect. There was, at last, an utter stillness, a waiting as though for a cosmic judgment. The eye, the torn eye, considered me.

"But I *do* love the world," I whispered to a waiting presence in the empty room. "I love its small ones, the things beaten in the strangling surf, the bird, singing, which flies and falls and is not seen again." I choked and said, with the torn eye still upon me, "I love the lost ones, the failures of the world." It was like the renunciation of my scientific heritage. The torn eye surveyed me sadly and was gone. I had come full upon one of the last great rifts in nature, and the merciless beam no longer was in traverse around my skull.

But no, it was not a rift but a joining: the expression of love projected beyond the species boundary by a creature born of Darwinian struggle, in the silent war under the tangled bank. "There is no boon in nature," one of the new philosophers had written harshly in the first years of the industrial cities. Nevertheless, through war and famine and death, a sparse mercy had persisted, like a mutation whose time had not yet come. I had seen the star thrower cross that rift and, in so doing, he had reasserted the human right to define his own frontier. He had moved to the utmost edge of natural being, if not across its boundaries. It was as though at some point the supernatural had touched hesitantly, for an instant, upon the natural.

Out of the depths of a seemingly empty universe had grown an eye, like the eye in my room, but an eye on a vastly larger scale. It looked out upon what I can only call itself. It searched the skies and it searched the depths of being. In the shape of man it had ascended like a vaporous emanation from the depths of night. The nothing had miraculously gazed upon the nothing and was not content. It was an intrusion into, or a projection out of, nature for which no precedent existed. The act was, in short, an assertion of value arisen from the domain of absolute zero. A little whirlwind of commingling molecules had succeeded in confronting its own universe.

Here, at last, was the rift that lay beyond Darwin's

tangled bank. For a creature, arisen from that bank and born of its contentions, had stretched out its hand in pity. Some ancient, inexhaustible, and patient intelligence, lying dispersed in the planetary fields of force or amidst the inconceivable cold of interstellar space, had chosen to endow its desolation with an apparition as mysterious as itself. The fate of man is to be the ever recurrent, reproachful Eye floating upon night and solitude. The world cannot be said to exist save by the interposition of that inward eye—an eye various and not under the restraints to be apprehended from what is vulgarly called the natural.

I had been unbelieving. I had walked away from the star thrower in the hardened indifference of maturity. But thought mediated by the eye is one of nature's infinite disguises. Belatedly, I arose with a solitary mission. I set forth in an effort to find the star thrower.

4

Man is himself, like the universe he inhabits, like the demoniacal stirrings of the ooze from which he sprang, a tale of desolations. He walks in his mind from birth to death the long resounding shores of endless disillusionment. Finally, the commitment to life departs or turns to bitterness. But out of such desolation emerges the awesome freedom to choose—to choose beyond the narrowly circumscribed circle that delimits the animal being. In that widening ring of human choice, chaos and order renew their symbolic struggle in the role of titans. They contend for the destiny of a world.

Somewhere far up the coast wandered the star thrower beneath his rainbow. Our exchange had been brief because upon that coast I had learned that men who ventured out at dawn resented others in the greediness of their compulsive collecting. I had also been abrupt because I had, in the terms of my profession and experience, nothing to say. The star thrower was mad, and his particular acts were a folly with which I had not chosen to associate myself. I was an observer and a scientist. Nevertheless, I had seen the rainbow attempting to attach itself to earth.

On a point of land, as though projecting into a domain beyond us, I found the star thrower. In the sweet rain-swept morning, that great many-hued rainbow still lurked and wavered tentatively beyond him. Silently I sought and picked up a still-living star, spinning it far out into the waves. I spoke once briefly. "I understand," I said. "Call me another thrower." Only then I allowed myself to think, He is not alone any longer. After us there will be others.

We were part of the rainbow—an unexplained projection into the natural. As I went down the beach I could feel the drawing of a circle in men's minds, like that lowering, shifting realm of color in which the thrower labored. It was a visible model of something toward which man's mind had striven, the circle of perfection.

I picked and flung another star. Perhaps far outward

on the rim of space a genuine star was similarly seized and flung. I could feel the movement in my body. It was like a sowing—the sowing of life on an infinitely gigantic scale. I looked back across my shoulder. Small and dark against the receding rainbow, the star thrower stooped and flung once more. I never looked again. The task we had assumed was too immense for gazing. I flung and flung again while all about us roared the insatiable waters of death.

But we, pale and alone and small in that immensity, hurled back the living stars. Somewhere far off, across bottomless abysses, I felt as though another world was flung more joyfully. I could have thrown in a frenzy of joy, but I set my shoulders and cast, as the thrower in the rainbow cast, slowly, deliberately, and well. The task was not to be assumed lightly, for it was men as well as starfish that we sought to save. For a moment, we cast on an infinite beach together beside an unknown hurler of suns. It was, unsought, the destiny of my kind since the rituals of the ice age hunters, when life in the Northern Hemisphere had come close to vanishing. We had lost our way, I thought, but we had kept, some of us, the memory of the perfect circle of compassion from life to death and back again to life—the completion of the rainbow of existence. Even the hunters in the snow, making obeisance to the souls of the hunted, had known the cycle. The legend had come down and lingered that he who gained the gratitude of animals gained help in need from the dark wood.

I cast again with an increasingly remembered sowing motion and went my lone way up the beaches. Somewhere, I felt, in a great atavistic surge of feeling, somewhere the Thrower knew. Perhaps he smiled and cast once more into the boundless pit of darkness. Perhaps he, too, was lonely, and the end toward which he labored remained hidden—even as with ourselves.

I picked up a star whose tube feet ventured timidly among my fingers while, like a true star, it cried soundlessly for life. I saw it with an unaccustomed clarity and cast far out. With it, I flung myself as forfeit, for the first time, into some unknown dimension of existence. From Darwin's tangled bank of unceasing struggle, selfishness, and death, had arisen, incomprehensibly, the thrower who loved not man, but life. It was the subtle cleft in nature before which biological thinking had faltered. We had reached the last shore of an invisible island—yet, strangely, also a shore that the primitives had always known. They had sensed intuitively that man cannot exist spiritually without life, his brother, even if he slays. Somewhere, my thought persisted, there is a hurler of stars, and he walks, because he chooses, always in desolation, but not in defeat.

In the night the gas flames under the shelling kettles would continue to glow. I set my clock accordingly. Tomorrow I would walk in the storm. I would walk against the shell collectors and the flames. I would walk remembering Bacon's forgotten words "for the uses of life." I would walk with the knowledge of the discontinuities of the unexpected universe. I would walk knowing of the rift revealed by the thrower, a hint that there looms, inexplicably, in nature something above the role men give her. I knew it from the man at the foot of the rainbow, the starfish thrower on the beaches of Costabel.

FIVE

The Angry Winter

As to what happened next, it is possible to maintain that the hand of heaven was involved, and also possible to say that when men are desperate no one can stand up to them. —XENOPHON

A time comes when creatures whose destinies have crossed somewhere in the remote past are forced to appraise each other as though they were total strangers. I had been huddled beside the fire one winter night, with the wind prowling outside and shaking the windows. The big shepherd dog on the hearth before me occasionally glanced up affectionately, sighed, and slept. I was working, actually, amidst the debris of a far greater winter. On my desk lay the lance points of ice age hunters and the heavy leg bone of a fossil bison. No remnants of flesh attached to these relics. The deed lay more than ten thousand years remote. It was represented here by naked flint and by bone so mineralized it rang when struck. As I worked on in my little circle of light, I absently laid the bone beside me on the floor. The hour had crept toward midnight. A grating noise, a heavy rasping of big teeth diverted me. I looked down.

The dog had risen. That rock-hard fragment of a vanished beast was in his jaws and he was mouthing it with a fierce intensity I had never seen exhibited by him before.

"Wolf," I exclaimed, and stretched out my hand. The dog backed up but did not yield. A low and steady rumbling began to rise in his chest, something out of a long-gone midnight. There was nothing in that bone to taste, but ancient shapes were moving in his mind and determining his utterance. Only fools gave up bones. He was warning me.

"Wolf," I chided again.

As I advanced, his teeth showed and his mouth wrinkled to strike. The rumbling rose to a direct snarl. His flat head swayed low and wickedly as a reptile's above the floor. I was the most loved object in his universe, but the past was fully alive in him now. Its shadows were whispering in his mind. I knew he was not bluffing. If I made another step he would strike.

Yet his eyes were strained and desperate. "Do not," something pleaded in the back of them, some affectionate thing that had followed at my heel all the days of his mortal life, "do not force me. I am what I am and cannot be otherwise because of the shadows. Do not reach out. You are a man, and my very god. I love you, but do not put out your hand. It is midnight. We are in another time, in the snow."

"The *other* time," the steady rumbling continued while I paused, "the other time in the snow, the big, the final, the terrible snow, when the shape of this thing I hold spelled life. I will not give it up. I cannot. The shadows will not permit me. Do not put out your hand."

I stood silent, looking into his eyes, and heard his whisper through. Slowly I drew back in understanding. The snarl diminished, ceased. As I retreated, the bone slumped to the floor. He placed a paw upon it, warningly.

And were there no shadows in my own mind, I wondered. Had I not for a moment, in the grip of that savage utterance, been about to respond, to hurl myself upon him over an invisible haunch ten thousand years removed? Even to me the shadows had whispered—to me, the scholar in his study.

"Wolf," I said, but this time, holding a familiar leash, I spoke from the door indifferently. "A walk in the snow." Instantly from his eyes that other visitant receded. The bone was left lying. He came eagerly to my side, accepting the leash and taking it in his mouth as always.

A blizzard was raging when we went out, but he paid no heed. On his thick fur the driving snow was soon clinging heavily. He frolicked a little—though usually he was a grave dog—making up to me for something still receding in his mind. I felt the snowflakes fall upon my face, and stood thinking of another time, and another time still, until I was moving from midnight to midnight under ever more remote and vaster snows. Wolf came to my side with a little whimper. It was he who was civilized now. "Come back to the fire," he nudged gently, "or you will be lost." Automatically I took the leash he offered. He led me safely home and into the house.

"We have been very far away," I told him solemnly. "I think there is something in us that we had both better try to forget." Sprawled on the rug, Wolf made no response except to thump his tail feebly out of courtesy. Already he was mostly asleep and dreaming. By the movement of his feet I could see he was running far upon some errand in which I played no part.

Softly I picked up his bone—our bone, rather—and replaced it high on a shelf in my cabinet. As I snapped off the light the white glow from the window seemed to augment itself and shine with a deep, glacial blue. As far as I could see, nothing moved in the long aisles of my neighbor's woods. There was no visible track, and certainly no sound from the living. The snow continued to fall steadily, but the wind, and the shadows it had brought, had vanished.

2

Vast desolation and a kind of absence in nature invite the emergence of equally strange beings or spectacular natural events. An influx of power accompanies nature's every hesitation; each pause is succeeded by an uncanny resurrection. The evolution of a lifeless planet eventually culminates in green leaves. The altered and oxygenated air hanging above the continents presently invites the rise of animal apparitions compounded of formerly inert clay.

Only after long observation does the sophisticated eye succeed in labeling these events as natural rather than miraculous. There frequently lingers about them a penumbral air of mystery not easily dispersed. We seem to know much, yet we frequently find ourselves baffled. Humanity itself constitutes such a mystery, for our species arose and spread in a time of great extinctions. We are the final product of the Pleistocene period's millennial winters, whose origin is still debated. Our knowledge of this ice age is only a little over a century old, and the time of its complete acceptance even less. Illiterate man has lost the memory of that huge snowfall from whose depths he has emerged blinking.

"Nature is a wizard," Thoreau once said. The self-styled inspector of snowstorms stood in awe of the six-pointed perfection of a snowflake. The air, even thin air, was full of genius. The poetic naturalist to whom, in our new-found scientism, we grudgingly accord a literary name but whom we dismiss as an indifferent investigator, made a profound observation about man during a moment of shivering thought on a frozen river. "The human brain," meditated the snowbound philosopher, "is the kernel which the winter itself matures." The winter, he thought, tended to concentrate and extend the power of the human mind.

"The winter," Thoreau continued, "is thrown to us like a bone to a famishing dog, and we are expected to get the marrow out of it." In foreshortened perspective Thoreau thus symbolically prefigured man's passage through the four long glacial seasons, from which we have indeed painfully learned to extract the marrow. Although Thoreau had seen the scratches left by the moving ice across Mount Monadnock, even to recording their direction, he was innocent of their significance. What he felt was a sign of his intuitive powers alone. He sensed uncannily the opening of a damp door in a remote forest, and he protested that nature was too big for him, that it was, in reality, a playground of giants.

Nor was Thoreau wrong. Man is the product of a very unusual epoch in earth's history, a time when the claws of a vast dragon, the glacial ice, groped fumbling toward him across a third of the world's land surface and blew upon him the breath of an enormous winter. It was a world of elemental extravagance, assigned by authorities to scarcely one per cent of earth's history and labeled "geo-catastrophic." For over a million years man, originally a tropical orphan, has wandered through age-long snowdrifts or been deluged by equally giant rains.

He has been present at the birth of mountains. He has witnessed the disappearance of whole orders of life and survived the cyclonic dust clouds that blew in the glacial winds off the receding ice fronts. In the end it is no wonder that he himself has retained a modicum of that violence and unpredictability which lie sleeping in the heart of nature. Modern man, for all his developed powers and his imagined insulation in his cities, still lives at the mercy of those giant forces that created him and can equally decree his departure. These forces are revealed in man's simplest stories—the stories in which the orphaned and abused prince evades all obstacles and, through the assistance of some benign sorcerer, slays the dragon and enters into his patrimony.

As the ice age presents a kind of caricature or sudden concentration of those natural forces that normally govern the world, so man, in the development of that awful instrument, his brain, himself partakes of the same qualities. Both his early magic and his latest science have magnified and frequently distorted the powers of the natural world, stirring its capricious and evil qualities. The explosive force of suns, once safely locked in nature, now lies in the hand that long ago dropped from a tree limb into the upland grass.

We have become planet changers and the decimators of life, including our own. The sorcerer's gift of fire in a dark cave has brought us more than a simple kingdom. Like so many magical gifts it has conjured up that which cannot be subdued but henceforth demands unceasing attention lest it destroy us. We are the genuine offspring of the sleeping ice, and we have inherited its power to magnify the merely usual into the colossal. The nature we have known throughout our venture upon earth has not been the stable, drowsy summer of the slow reptilian ages. Instead, we are the final product of a seemingly returning cycle, which comes once in two hundred and fifty million years—about the time, it has been estimated, that it takes our sun to make one full circle of the galactic wheel.

That circle and its recurrent ice have been repeated back into dim pre-Cambrian eras, whose life is lost to us. When our first tentative knowledge of the cold begins, the time is Permian. This glaciation, so far as we can determine, was, in contrast to the ice age just past, confined primarily to the southern hemisphere. Like the later Pleistocene episode, which saw the rise of man, it was an epoch of continental uplift and of a steeper temperature gradient from the poles to the equator. It produced a crisis in the evolution of life that culminated in the final invasion of the land by the reptilian vertebrates. More significantly, so far as our own future was concerned, it involved the rise of those transitional twilight creatures, the mammal-like reptiles whose remote descendants we are.

They were moving in the direction of fur, warm blood, and controlled body temperature, which, in time, would give the true mammals the mastery of the planet. Their forerunners were the first vertebrate responses to the recurrent menace of the angry dragon. Yet so far away in the past, so dim and distant was the breath of that frosty era that the scientists of the nineteenth century, who believed in a constant heat loss from a once fiery earth, were amazed when A. C. Ramsay, in 1854, produced evidence that the last great winter, from which man is only now emerging, had been long preceded in the Permian by a period of equally formidable cold.

Since we live on the borders of the Pleistocene, an ice age that has regressed but not surely departed, it is perhaps well to observe that the older Permian glaciation is the only one of whose real duration we can form a reasonable estimate. Uncertain traces of other such eras are lost in ancient strata or buried deep in the pre-Cambrian shadows. For the Permian glaciation, however, we can derive a rough estimate of some twenty-five to thirty million years, during which the southern continents periodically lay in the grip of glacial ice. Philip King, of the U.S. Geological Survey, has observed that in Australia the period of Permian glaciation was prolonged, and that in eastern Australia boulder beds of glacial origin are interspersed through more than ten thousand feet of geological section. The temperature gradient of that era would never again be experienced until the onset of the cold that accompanied the birth of man.

If the cause of these glacial conditions, with the enormous intervals between them, is directed by recurrent terrestrial or cosmic conditions, then man, unknowingly, is huddling memoryless in the pale sunshine of an interstadial spring. Ice still lies upon the poles; the arctic owl, driven south on winter nights, drifts white and invisible over the muffled countryside. He is a survival from a vanished world, a denizen of the long cold of which he may yet be the returning harbinger.

I have said that the earlier Permian glaciation appears to have fluctuated over perhaps thirty million years. Some two hundred and fifty million years later, the Pleistocene—the ice age we call our own—along with four interglacial summers (if we include the present), has persisted a scant million years.* So recent is it that its two earlier phases yield little evidence of animal adjustment to the cold. "The origin of arctic life," remarks one authority, "is shrouded in darkness." Only the last two ice advances have given time, apparently, for the emergence of a fauna of arctic aspect —the woolly mammoths, white bears, and tundra-grazing reindeer, who shared with man the experience of the uttermost cold.

The arctic, in general, has been the grave of life rather than the place of its primary development. Man is the survivor among many cold-adapted creatures who streamed away at last with the melting glaciers. So far, the Pleistocene, in which geologically we still exist, has been a time of great extinctions. Its single new emergent, man, has himself contributed to making it what Alfred Russel Wallace has called a "zoologically impoverished world." Judging from the Permian record, if we were to experience thirty million more years of alternate ice and sun across a third of the earth's surface, man's temperate-zone cities would be ground to powder, his numbers decimated, and he himself might die in bestial savagery and want. Or, in his new-found scientific cleverness, he might survive his own unpre-

*Late discoveries have extended the Tertiary time range of the protohuman line. *Homo sapiens* may have existed for a time contemporaneously with the last of the heavy-browed forms of man, well back in the Pleistocene. If so, however, there is suggestive evidence that fertile genetic mixture between the two types existed. The human interminglings of hundreds of thousands of years of prehistory are not to be clarified by a single generation of archaeologists.

dictable violence and live on as an archaic relic, a dropped pebble from a longer geological drama.

Already our own kind, *Homo sapiens,* with the assistance of the last two ice advances, appears to have eliminated, directly or indirectly, a sizable proportion of the human family. One solitary, if fertile, species, lost in internecine conflict, confronts the future even now. Man's survival record, for all his brains, is not impressive against the cunning patience of the unexpired Great Winter. In fact, we would do well to consider the story of man's past and his kinship with the planetary dragon—for of this there is no doubt at all. "The association of unusual physical conditions with a crisis in evolution is not likely to be pure coincidence," George Gaylord Simpson, a leading paleontologist, has declared. "Life and its environment are interdependent and evolve together."

The steps to man begin before the ice age. Just as in the case of the ancestral mammals, however, they are heralded by the oncoming cold of the late Age of Mammals, the spread of grass, the skyward swing of the continents, and the violence of mountain upheaval. The Pleistocene episode, so long unguessed and as insignificant as a pinprick on the earth's great time scale, signifies also, as did the ice of the late Paleozoic, the rise of a new organic world. In this case it marked in polycentric waves, distorted and originally hurled back by the frost, the rise of a creature not only new, but also one whose head contained its own interior lights and shadows and who was destined to reflect the turbulence and beauty of the age in which it was born.

It was an age in which the earth, over a third of its surface, over millions of square miles of the Northern Hemisphere, wore a mantle of blue ice stolen from the shrinking seas. And as that mantle encased and covered the final strata of earth, so, in the brain of man, a similar superimposed layer of crystalline thought substance superseded the dark, forgetful pathways of the animal brain. Sounds had their origin there, strange sounds that took on meaning in the air, named stones and gods. For the first time in the history of the planet, living men received names. For the first time, also, men wept bitterly over the bodies of their dead.

There was no longer a single generation, which bred blindly and without question. Time and its agonizing nostalgia would touch the heart each season and be seen in the fall of a leaf. Or, most terrible of all, a loved face would grow old. Kronos and the fates had entered into man's thinking. Try to escape as he might, he would endure an interior ice age. He would devise and unmake fables and at last, and unwillingly, comprehend an intangible abstraction called space-time and shiver inwardly before the endless abysses of space as he had once shivered unclothed and unlighted before the earthly frost.

As Thoreau anticipated, man has been matured by winter; he has survived its coming, and has eaten of its marrow. But its cold is in his bones. The child will partake always of the parent, and that parent is the sleeping dragon whose kingdom we hold merely upon sufferance, and whose vagaries we have yet to endure.

3

A few days ago I chanced to look into a rain pool on the walk outside my window. For a long time, because I was dreaming of other things, I saw only the occasional spreading ripple from a raindrop or the simultaneous overlapping circles as the rain fell faster. Then, as the beauty and the strange rhythm of the extending and concentric wavelets entered my mind, I saw that I was looking symbolically upon the whole history of life upon our globe. There, in a wide, sweeping circle, ran the early primates from whom we are descended; here, as a later drop within the rim of the greater circle, emerged the first men. I saw the mammoths pass in a long, slow, world-wide surge, but the little drop of man changed into a great hasty wave that swept them under.

There were sudden little ringlets, like the fauna of isolated islands, that appeared and disappeared with rapidity. Sometimes so slow were the drops that the pool was almost quiet, like the intense, straining silence of a quiescent geological period. Sometimes the rain, like the mutations in animal form, came so fast that the ripples broke, mixed, or kept their shapes with difficulty and did not spread far. Jungles, I read in my mystical water glass, microfaunas changing rapidly but with little spread.

Watch instead, I thought, for the great tides—it is they that contain the planet's story. As the rain hastened or dripped slowly, the pictures in the little pool were taken into my mind as though from the globe of a crystal-gazer. How often, if we learn to look, is a spider's wheel a universe, or a swarm of summer midges a galaxy, or a canyon a backward glance into time. Beneath our feet is the scratched pebble that denotes an ice age, or above us the summer cloud that changes form in one afternoon as an animal might do in ten million windy years.

All of these perceptive insights that we obtain from the natural world around us depend upon painfully accumulated knowledge. Otherwise, much as to our ancestors, the pebble remains a pebble, the pool but splashing water, the canyon a deep hole in the ground. Increasingly, the truly perceptive man must know that where the human eye stops, and hearing terminates, there still vibrates an inconceivable and spectral world of which we learn only through devised instruments. Through such instruments measuring atomic decay we have learned to probe the depths of time before our coming and to gauge temperatures long vanished.

Little by little, the orders of life that had characterized the earlier Age of Mammals ebbed away before the oncoming cold of the Pleistocene, interspersed though this cold was by interglacial recessions and the particularly long summer of the second interglacial. There were times when ice accumulated over Britain; in the New World, there were times when it stretched across the whole of Canada and reached southward to the fortieth parallel of latitude, in what today would be Kansas.

Manhattan Island and New Jersey felt its weight. The giant, long-horned bison of the middle Pleistocene vanished before man had entered America. Other now extinct but less colossal forms followed them. By the closing Pleistocene, it has been estimated, some seventy per cent of the animal life of the Western world had perished. Even in Africa, remote from the ice centers, change was evident. Perhaps the devastation was a partial response to the Pluvials, the great rains that in the tropics seem to have accompanied or succeeded the ice advances in the north.

The human groups that existed on the Old World land mass were alternately squeezed southward by advancing ice, contracted into pockets, or released once more to find their way northward. Between the alternate tick and tock of ice and sun, man's very bones were changing. Old species passed slowly away in obscure refuges or fell before the weapons devised by sharper minds under more desperate circumstances. Perhaps, since the rise of mammals, life had been subjected to no more drastic harassment, no more cutting selective edge, no greater isolation and then renewed genetic commingling. Yet we know that something approximating man was on the ground before the ice commenced and that naked man is tropical in origin. What, then, has ice to do with his story?

It has, in fact, everything. The oncoming chill caught him early in his career; its forces converged upon him even in the tropics; its influence can be seen in the successive human waves that edge farther and ever farther north until at last they spill across the high latitudes at Bering Strait and descend the length of the two Americas. Only then did the last southwestern mammoths perish in the shallow mud of declining lakes, the last mastodons drop their tired bones in the New Jersey bogs on the receding drift.

The story can best be seen from the map, as time, ice, and the sorcerer's gift of fire run like the concentric ripples of the falling rain across the zones of temperature. The tale is not confined to ice alone. As one glaciologist, J. K. Charlesworth, has written: "The Pleistocene . . . witnessed earth-movements on a considerable, even catastrophic scale. There is evidence that it created mountains and ocean deeps of a size previously unequalled. . . . The Pleistocene represents one of the crescendi in the earth's tectonic history."

I have spoken of the fact that, save for violent glacial episodes, the world's climate has been genial. The planet has been warmer than today—"acryogenic," as the specialists would say. Both earlier and later, warm faunas reached within the Arctic Circle, and a much higher percentage of that fauna represented forest forms. Then in the ice phases, world temperature dipped, even in the tropics; the mountain snowlines crept downward. In the north, summers were "short and false," periods of "dry cold"—again to quote the specialists. Snow blanketed the high ground in winter, and that winter covered half the year and was extremely harsh.

With our short memory, we accept the present climate as normal. It is as though a man with a huge volume of a thousand pages before him—in reality, the pages of earth time—should read the final sentence on the last page and pronounce it history. The ice has receded, it is true, but world climate has not completely rebounded. We are still on the steep edge of winter or early spring. Temperature has reached mid-point. Like refugees, we have been dozing memoryless for a few scant millennia before the windbreak of a sun-warmed rock. In the European Lapland winter that once obtained as far south as Britain, the temperature lay eighteen degrees Fahrenheit lower than today.

On a world-wide scale this cold did not arrive unheralded. Somewhere in the highlands of Africa and Asia the long Tertiary descent of temperature began. It was, in retrospect, the prelude to the ice. One can trace its presence in the spread of grasslands and the disappearance over many areas of the old forest browsers. The continents were rising. We know that by Pliocene time, in which the trail of man ebbs away into the grass, man's history is more complicated than the simple late descent, as our Victorian forerunners sometimes assumed, of a chimpanzee from a tree. The story is one whose complications we have yet to unravel.

Avoiding complexities and adhering as we can to our rain-pool analogy, man, subman, protoman, the euhominid, as we variously denote him, was already walking upright on the African grasslands more than two million years ago. He appears smaller than modern man, pygmoid and light of limb. Giantism comes late in the history of a type and sometimes foretells extinction. Man is now a giant primate and, where food is plentiful, growing larger, but he is a unique creature whose end is not yet foreseeable.

Three facts can be discerned as we examine the earliest bipedal man-apes known to us. First, they suggest, in their varied dentition and skull structure, a physical diversity implying, as Alfred Russel Wallace theorized long ago, an approach to that vanished era in which protoman was still being molded by natural selective processes unmediated by the softening effect of cultural defenses. He was, in other words, scant in numbers and still responding genetically to more than one ecological niche. Heat and cold were direct realities; hunger drove him, and, on the open savanna into which he had descended, vigilance was the price of life. The teeth of the great carnivores lay in wait for the old, the young, and the unwary.

Second, at the time we encounter man, the long descent of the world's climate toward the oncoming Pleistocene cold had already begun. It is not without interest that all man's most primitive surviving relatives—living fossils, we would call them—are tree dwellers hidden in the tropical rain forests of Africa, Madagascar, and the islands of southeastern Asia. They are the survivors of an older and a warmer world—the incubation time that was finally to produce, in some unknown fashion, the world-encircling coils of the ice dragon.

In the last of Tertiary time, grasslands and high country were spreading even in the tropics. Savanna parkland interspersed with trees clothed the uplands of East Africa; North China grew colder and more arid. Steppe- and plains-loving animals became predominant. Even the seas grew colder, and the tropical zone narrowed. Africa was to remain the least glaciated of the continents, but even here the lowering of temperature drew on, and the mountain glaciations finally began to descend into their valleys. As for Asia, the slow, giant upthrust of the Himalayas had brought with it the disappearance of jungles harboring the old-fashioned tree climbers.

Of the known regions of late-Tertiary primate development, whether African or Asian, both present the spectacle of increasing grasslands and diminishing forest. The latter, as in southeastern Asia, offered a refuge for the arboreal conservatives, such as the gibbon and orang, but the Miocene-Pliocene parklands and savannas must have proved an increasing temptation to an intelligent anthropoid sufficiently unspecialized and agile to venture out upon the grass. Our evidence from Africa is more complete at present, but fragmentary remains that may prove to be those of equally bipedal creatures are known from pre-Pleistocene and less explored regions below the Himalayas.

Third, and last of the points to be touched on here, the man-apes, in venturing out erect upon the grass, were leaving forever the safety of little fruit-filled niches in the forest. They were entering the open sunlight of a one-dimensional world, but they were bringing to that adventure a freed forelimb at the conscious command of the brain, and an eye skillfully adjusted for depth perception. Increasingly they would feed on the rich proteins provided by the game of the grasslands; by voice and primitive projectile weapon, man would eventually become a space leaper more deadly than the giant cats.

In the long, chill breath that presaged the stirring of the world dragon, the submen drifted naked through an autumnal haze. They were, in body, partly the slumbering product of the earth's long summer. The tropical heat had warmed their bones. Thin-furred and hungry, old-fashioned descendants from the forest attic, they clung to the tropical savannas. Unlike the light gazelle, they could neither bound from enemies nor graze on the harsh siliceous grasses. With a minimum of fragmentary chips and stones, and through an intensified group co-operation, they survived.

The first human wave, however, was a little wave, threatening to vanish. A patch in Africa, a hint in the Siwalik beds below the Himalayas—little more. Tremendous bodily adjustments were in process, and, in the low skull vault, a dream animal was in the process of development, a user of invisible symbols. In its beginnings, and ever more desperately, such a being walks the knife-edge of extinction. For a creature who dreams outside of nature, but is at the same time imprisoned within reality, has acquired, in the words of the psychiatrist Leonard Sillman, "one of the cruelest and most generous endowments ever given to a species of life by a mysterious providence."

On that one most recent page of life from which we can still read, it is plain that the second wave of man ran onward into the coming of the ice. In China a pithecanthropine creature with a cranial capacity of some 780 cubic centimeters has been recently retrieved from deposits suggesting a warm grassland fauna of the lower Pleistocene, perhaps over 700,000 years remote from the present. The site lies in Shensi province in about thirty degrees north latitude. Man is moving northward. His brain has grown, but he appears still to lack fire.

4

In the legendary cycles of the Blackfoot Indians there is an account of the early people, who were poor and naked and did not know how to live. Old Man, their maker, said: "Go to sleep and get power. Whatever animals appear in your dream, pray and listen." "And," the story concludes, "that was how the first people got through the world, by the power of their dreams."

Man was not alone young and ignorant in the morning of his world; he also died young. Much of what he grasped of the world around him he learned like a child from what he imagined, or was gleaned from his own childlike parents. The remarks of Old Man, though clothed in myth, have an elemental ring. They tell the story of an orphan—man—bereft of instinctive instruction and dependent upon dream, upon, in the end, his own interpretation of the world. He had to seek animal helpers because they alone remembered what was to be done.

And so the cold gathered and man huddled, dreaming, in the lightless dark. Lightning struck, the living fire ran from volcanoes in the fury of earth's changes, and still man slumbered. Twice the ice ground southward and once withdrew, but no fire glimmered at a cave mouth. Humanly flaked flints were heavier and better-made. Behind that simple observation lies the unknown history of drifting generations, the children of the dreamtime.

At about the forty-fifth parallel of latitude, in the cave vaults at Choukoutien, near Peking, a heavy-browed, paleoanthropic form of man with a cranial capacity as low, in some instances, as 860 cubic centimeters, gnawed marrow bones and chipped stone implements. The time lies 500,000 years remote; the hour is late within the second cold, the place northward and more bleak than Shensi.

One thing strikes us immediately. This creature, with scarcely two-thirds of modern man's cranial capacity, was a fire user. Of what it meant to him beyond warmth and shelter we know nothing; with what rites, ghastly or benighted, it was struck or maintained, no word remains. We know only that fire opened to man the final conquest of the earth.

I do not include language here, in spite of its tremendous significance to humanity, because the potentiality of language is dependent upon the germ plasm. Its nature, not its cultural expression, is written into the motor centers of the brain, into high auditory discrimination and equally rapid neuromuscular response in tongue, lips, and palate. We are biologically adapted for the symbols of speech. We have determined its forms, but its potential is not of our conscious creation. Its mechanisms are written in our brain, a simple gift from the dark powers behind nature. Speech has made us, but it is a human endowment not entirely of our conscious devising.

By contrast, the first fires flickering at a cave mouth are our own discovery, our own triumph, our grasp upon invisible chemical power. Fire contained, in that place of brutal darkness and leaping shadows, the crucible and the chemical retort, steam and industry. It contained the entire human future.

Across the width of the Old World land mass near what is now Swanscombe, England, a better-brained creature of almost similar dating is also suspected of using fire, though the evidence, being from the open, is not so clear. But at last the sorcerer-priest, the stealer from the gods, the unknown benefactor remembered in a myriad legends around the earth, had done his work. He had supplied man with an overmastering magic. It would stand against the darkness and the cold.

In the frontal and temporal lobes, anatomy informs us, lie areas involved with abstract thought. In modern man the temporal lobes in particular are "hazardously supplied with blood through tenuous arteries. They are protected by a thin skull and crowded against a shelf of bone. They are more commonly injured than any other higher centers." The neurologist Frederick Gibbs goes on to observe that these lobes are attached to the brain like dormer windows, jammed on as an afterthought of nature. In the massive armored cranium of

Peking man those lobes had already lit the fires that would knit family ties closer, promote the more rapid assimilation of wild food, and increase the foresight that goes into the tending of fires always. Fire is the only natural force on the planet that can both feed and travel. It is strangely like an animal; that is, it has to be tended and fed. Moreover, it can also rage out of control.

Man, long before he trained the first dog, had learned to domesticate fire. Its dancing midnight shadows and the comfort it gave undoubtedly enhanced the opportunities for brain growth. The fourth ice would see man better clothed and warmed. In our own guise, as the third and last great human wave, man would pursue the trail of mammoths across the Arctic Circle into America. The animal counselors that once filled his dreams would go down before him. Thus, inexorably, he would be forced into a new and profound relationship with plants. If one judges by the measures of civilization, it was all for the best. There are, however, lingering legends that carry a pathetic symbolism: that it was fire that separated man from the animals. It is perhaps a last wistful echo from a time when the chasm between ourselves and the rest of life did not yawn so impassably.

5

They tell an old tale in camping places, where men still live in the open among stones and trees. Always, in one way or another, the tale has to do with messages, messages that the gods have sent to men. The burden of the stories is always the same. Someone, man or animal, is laggard or gets the message wrong, and the error is impossible to correct; thus have illness and death intruded in the world.

Mostly the animals understand their roles, but man, by comparison, seems troubled by a message that, it is often said, he cannot quite remember, or has gotten wrong. Implied in this is our feeling that life demands an answer from us, that an essential part of man is his struggle to remember the meaning of the message with which he has been entrusted, that we are, in fact, message carriers. We are not what we seem. We have had a further instruction.

There is another story that is sometimes told of the creator in the morning of the world. After he had created the first two beings, which he pronounced to be "people," the woman, standing by the river, asked: "Shall we always live?" Now the god had not considered this, but he was not unwilling to grant his new creations immortality. The woman picked up a stone and, gesturing toward the stream, said: "If it floats we shall always live, but if it sinks, people must die so that they shall feel pity and have compassion." She tossed the stone. It sank. "You have chosen," said the creator.

Many years ago, as a solitary youth much given to wandering, I set forth on a sullen November day for a long walk that would end among the fallen stones of a forgotten pioneer cemetery on the High Plains. The weather was threatening, and only an unusual restlessness drove me into the endeavor. Snow was on the ground and deepening by the hour. There was a rising wind of blizzard proportions sweeping across the land.

Late in a snow-filled twilight, I reached the cemetery. The community that placed it there had long vanished. Frost and snow, season by season, had cracked and shattered the flat, illegible stones till none remained upright. It was as though I, the last living man, stood freezing among the dead. I leaned across a post and wiped the snow from my eyes.

It was then I saw him—the only other living thing in that bleak countryside. We looked at each other. We had both come across a way so immense that neither my immediate journey nor his seemed of the slightest importance. We had each passed over some immeasurably greater distance, but whatever the word we had carried, it had been forgotten between us.

He was nothing more than a western jack rabbit, and his ribs were gaunt with hunger beneath his skin. Only the storm contained us equally. That shrinking, long-eared animal, cowering beside a slab in an abandoned graveyard, helplessly expected the flash of momentary death, but it did not run. And I, with the rifle so frequently carried in that day and time, I also stood while the storm—a real blizzard now—raged over and between us, but I did not fire.

We both had a fatal power to multiply, the thought flashed on me, and the planet was not large. Why was it so, and what was the message that somehow seemed spoken from a long way off beyond an ice field, out of all possible human hearing?

The snow lifted and swirled between us once more. He was going to need that broken bit of shelter. The temperature was falling. For his frightened, trembling body in all the million years between us, there had been no sorcerer's aid. He had survived alone in the blue nights and the howling dark. He was thin and crumpled and small.

Step by step I drew back among the dead and their fallen stones. Somewhere, if I could follow the fence lines, there would be a fire for me. For a moment I could see his ears nervously recording my movements, but I was a wraith now, fading in the storm.

"There are so few tracks in all this snow," someone had once protested. It was true. I stood in the falling flakes and pondered it. Even my own tracks were filling. But out of such desolation had arisen man, the desolate. In essence, he is a belated phantom of the angry winter. He carried, and perhaps will always carry, its cruelty and its springtime in his heart.

SIX

The Golden Alphabet

A creature without memory cannot discover the past; one without expectation cannot conceive a future.

—GEORGE SANTAYANA

"WISDOM," the Eskimo say, "can be found only far from man, out in the great loneliness." These people speak from silences we will not know again until we set foot upon the moon. Perhaps our track is somehow rounding evocatively backward into another version of the giant winter out of which we emerged ten thousand years ago. Perhaps it is our destiny to have plunged across it only to re-enter it once more.

Of all the men of the nineteenth century who might be said to have been intimates of that loneliness and yet, at the same time, to have possessed unusual prophetic powers, Henry David Thoreau and Charles Darwin form both a spectacular comparison and a contrast. Both Thoreau and Darwin were voyagers. One confined himself to the ever widening ripples on a pond until they embraced infinity. The other went around the world and remained for the rest of his life a meditative recluse in an old Victorian house in the English countryside.

The two men shared a passion for odd facts. In much else they differed. Darwin, after long travel, had immured himself at home. Thoreau could only briefly tolerate a dwelling, and his journals suggest that he suffered from claustrophobic feelings that a house was a disguised tomb, from which he had constantly to escape into the open. "There is no circulation there," he once protested.

Both men were insatiable readers and composers of works not completely published in their individual lifetimes. Both achieved a passionate satisfaction out of their association with the wilderness. Each in his individual way has profoundly influenced the lives of the generations that followed him. Darwin achieved fame through a great biological synthesis—what Thoreau would have called the demoniacal quality of the man who can discern a law, or couple two facts. Thoreau, by contrast, is known as much for what he implied as for what he spoke. His life, like Darwin's, is known but in many ways hidden. As he himself intimated cryptically, he had long ago lost a bay horse, a hound dog, and a turtledove, for which he was searching. It is not known that he ever came upon them or precisely what they represented.

All his life Thoreau dwelt along the edge of that visible nature of which Darwin assumed the practical mastery. Like the owls Thoreau described in *Walden,* he himself represented the stark twilight of a nature "behind the ordinary," which has passed unrecognized. As he phrased it, "We live on the outskirts of that region. . . . We are still being born, and have as yet but a dim vision."

Both men forfeited the orthodox hopes that had sustained, through many centuries, the Christian world. Yet, at the last, the one transcends the other's vision, or amplifies it. Darwin remains, though sometimes hesitantly, the pragmatic scientist, content with what his eyes have seen. The other turns toward an unseen spring beyond the wintry industrialism of the nineteenth century, with its illusions of secular progress. The two views, even the two lives, can be best epitomized in youthful expressions that have come down to us. The one, Darwin's, is sure, practical, and exuberant. The other reveals an exploring, but wary, nature.

Darwin, the empiricist, wrote from Valparaiso in 1834: "I have just got scent of some fossil bones of a MAMMOTH; what they may be I do not know, but if gold or galloping will get them they shall be mine." Thoreau, by nature more skeptical of what can be captured in this world, mused, in his turn, "I cannot lean so hard on any arm as on a sunbeam." It was one of the first of many similar enigmatic expressions that were finally to lead his well-meaning friend, Ellery Channing, to venture sadly, "I have never been able to understand what he meant by his life. . . . Why was he so disappointed with everybody else? Why was he so interested in the river and the woods . . . ? Something peculiar here I judge."

Channing was not wrong. There *was* something peculiar about Thoreau, just as there was something equally peculiar about Darwin. The difference between them lies essentially in the nature of man himself, the creature who persists in drawing sharp, definitive lines across the indeterminate face of nature. Essentially, the problem may be easily put. It is its varied permutations and combinations that each generation finds so defeating, and that our own time is busy, one might say horribly busy, in re-creating.

One may begin with what we all remember from childhood—the emerald light in the wonderful city of Oz. Those who lived in the city wore spectacles that were locked on by night and day. Oz had so ordered it when the city was built. Now Oz, it was explained to the simple ones who came there, was a great wizard who could take on any form he wished. "But," as one denizen of the city explained, "who the real Oz is when he is in his own form, no living person can tell."

Among the visitors to that city came several creatures, only one of whom was human, but all of whom dealt with great questions couched in very simple form. One was the Tin Woodman, in search of a heart. One was the Cowardly Lion, who had not the courage to keep tramping forever without getting anywhere at all. Then there was also the little girl, Dorothy, from Kansas, who was sure that if they walked far enough they would sometime come to some place. Particularly pertinent here is that appealing character, the Scarecrow, who, with his straw-filled head and patient good nature has always represented the better, more humble side of man. Scarecrow had been made out of straw instead of the clay so frequently utilized in the creation of man, and perhaps he proved the better for it. At any rate, his only recorded comment upon his existence in the fields was, "It was a lonely life to lead, for I had nothing to think of, having been made such a little while before."

The whole story of humanity is basically that of a journey toward the Emerald City, and of an effort to learn the nature of Oz, who, perhaps wisely, keeps himself concealed. In each human heart exists the Cowardly Lion and the little girl who was sure that the solution to life lay in just walking far enough. Finally, among our great discoverers are those with precious straw-filled heads who have to make up their own thoughts because each knows he has been made such a little while before, and has stood alone in the fields. Darwin and Thoreau are two such oddly opposed, yet similar, scarecrows. As it turned out, they came to two different cities, or at least vistas. They discovered something of the nature of Oz, and, rightly understood, their views are complementary to each other.

I shall treat first of Darwin and then of Thoreau, because, though contemporaries, they were distinct in temperament. Thoreau, who died young, perhaps trudged farther toward the place which the little girl Dorothy was so sure existed, and thus, in a sense, he may be a messenger from the future. Since futures do not really exist until they are present, it might be more cautious to say that Thoreau was the messenger of a *possible* future in some way dependent upon ourselves.

Neither of the two men ever discovered the nature of Oz himself. The one, Darwin, learned much about his ways—so much, indeed, that I suspect he came to doubt the existence of Oz. The other, Thoreau, leaned perhaps too heavily upon his sunbeam, and in time it faded, but not surely, because to the last he clung to the fields and heard increasingly distant echoes. Both men wore spectacles of sorts, for this is a rule that Oz has decreed for all men. Moreover, there are diverse kinds of spectacles.

There are, for example, the two different pairs through which philosophers may look at the world. Through one we see ourselves in the light of the past; through the other, in the light of the future. If we fail to use both pairs of spectacles equally, our view of ourselves and of the world is apt to be distorted, since we can never see completely without the use of both. The historical sciences have made us very conscious of our past, and of the world as a machine generating successive events out of foregoing ones. For this reason some scholars tend to look totally backward in their interpretation of the human future. It is, unconsciously, an exercise much favored in our time.

Like much else, this attitude has a history.

When science, early in the nineteenth century, began to ask what we have previously termed "the terrible questions" because they involved the nature of evil, the age of the world, the origins of man, of sex, or even of language itself, a kind of Pandora's box had been opened. People could classify giraffes and porcupines but not explain them—much less a man. Everything stood in isolation, and therefore the universe of life was bound to appear a little ridiculous to the honestly enquiring mind. What was needed was the kind of man of whom Thoreau had spoken, who could couple two seemingly unrelated facts and reduce the intractable chaos of the world. Such a man was about to appear. In fact, he had already had his forerunners.

2

Robert Fitzroy was a captain with a conscience. Another of that great breed of English navigators of whom Cook stands as the epitome, Fitzroy had been appointed at twenty-three to the command of H.M.S. *Beagle* on a mapping and exploring voyage around Cape Horn. This event preceded the famous expedition in which Charles Darwin was to participate. Mapping the Strait of Magellan, the ambitious young officer discovered, was rather like mapping the stars in the heavens. Perhaps Fitzroy's wry comment to this effect was an unconscious omen of what was to prove the task of Charles Darwin, venturing upon the greater waters of time and change.

The second voyage might never have taken place had it not been that in his adventures about the Strait Fitzroy had acquired four savage Indians, whom he brought home to London in 1830 with the quixotic idea of familiarizing them with Christian civilization and then returning them to their native land. One man died of smallpox—the other three, one woman and two men, survived. The troubled captain, maintaining and attempting to educate these people at his own expense, decided upon their return, even if he personally had to charter a vessel to see them safely home.

Ironically, it was because of this touch of zealous missionary spirit on the part of Fitzroy that mankind was to find itself eventually displaced, biologically, from the center of the universe. The aristocratic Fitzroy exerted influence upon the Admiralty, and that body, in turn, assented to a second voyage, under renewed instructions for further mapping and exploration.

Fitzroy, at heart a lonely young captain, sought a companion. In the Cook tradition he decided upon a naturalist. Charles Darwin secured the post through the good offices of his botanical instructor at Cambridge, John Henslow. Fitzroy, a passionately religious early Victorian, had taken aboard his vessel a man who by training and inclination carried with him the liberal enquiring attitudes of the Enlightenment—the spirit that had perished in the excesses of the French Revolution.

The shadow of those excesses had fallen darkly across Britain and had accentuated the conservatism of the upper classes. There was a strong tendency to excoriate or ridicule French thinkers, particularly if their ideas appeared religiously unorthodox. Even their rare defenders preferred to remain anonymous. The result, for the historian of science, is unfortunate. Published innovations in some instances remain unidentifiable with their advocates. In other cases, anticipations of what were later to emerge as significant ideas have been hidden, deliberately or by accident, under innocuous titles, or interjected into seemingly guileless and innocent works upon stultifying subjects.

We know perfectly well, for example, that the name of the French evolutionist Jean Baptiste Lamarck was known in Edinburgh University circles when young Charles Darwin was a medical student there. We know, even, that one of Darwin's instructors, Robert Grant, was an avowed follower of the French naturalist. Yet so fixed was this isolated British set of mind that as late as 1863, after the publication of the *Origin of Species,* we find Darwin writing to Sir Charles Lyell on the subject of the *Philosophie Zoologique,* "to me it was an absolutely useless book," owing, he says, to his search for "facts."

In the same letter he dismisses his grandfather's ideas with equal abruptness. Not only are these remarks scarcely borne out by a careful examination of Darwin's work, but the harsh emphasis upon "facts" comes a little oddly from a scholar who could also protest, "Forfend me from a man who weighs every expression with Scotch prudence." Elsewhere he intimates he can scarcely abide facts without attempting to tie them together.

The youth who went aboard the *Beagle* in December 1831 was a great deal more clever than his academic record at both Edinburgh and Cambridge might suggest. The ingenuity with which he went about securing his father's permission for the voyage in itself indicates the dedicated persistence with which he could overcome obstacles. There remained in the motherless young man a certain wary reserve, which would finally draw him into total seclusion. In the first edition of the *Origin of Species* he was to write: "When on board H.M.S. *Beagle,* as naturalist, I was much struck with certain facts. . . . These facts seemed to . . . throw light on the origin of species. . . ." The remark is true, but it is also ingenuous. Young Charles's first knowledge of evolution did not emerge spontaneously aboard the *Beagle,* however much that conception was to be strengthened in the wild lands below the equator.

Instead, its genesis in Darwin's mind lies mysteriously back amidst unrecorded nights in student Edinburgh and lost in the tracery of spider tracks over thousands of dusty volumes after his return. For this is the secret of Charles Darwin the naturalist-voyager, the modern Odysseus who came to Circe's island of change in the Galápagos: he was the product of two odysseys, not one. He lives in the public mind partly by the undoubted drama of a great voyage whose purpose, as defined by the chief hydrographer, was the placing of a chain of meridians around the world. While those meridians were being established through Fitzroy's efforts, another set was being posted by Darwin in the haunted corridors of the past.

But the second odyssey, the one most solitary, secretive, and hidden, is the Merlin-like journey which had no ending save at death. It is the groping through webby corridors of books in smoky London—the kind of journey in which men are accountable only to themselves and by which the public is not at all enlivened. No waves burst, no seaman falls from the masthead, no icy continent confronts the voyager. Within the mind, however, all is different. There are ghost fires burning over swampy morasses of books, confusing trails, interceptions of the lost, the endless weaving and unweaving of floating threads of thought drawn from a thousand sources.

"Lord," briefly writes the man with the increasingly worn face and heavy brow, "in what a medley the origin of cultivated plants is. I have been reading on strawberries, and I can find hardly two botanists agree what are the wild forms; but I pick out horticultural books here and there, with queer cases of variation." Or, being the man he was, again surfacing for a moment from the least expected place: "I sat one evening in a gin palace among pigeon fanciers."

A doubt, a shrinking terrible doubt such as can cause a man's hands to shake and rustle amidst the leaves of folios, came on him in 1854. "How awfully flat I shall feel, if when I get my notes together . . . the whole thing explodes like an empty puff-ball."

References to old transactions, old travels, old gardener's magazines bestrew his letters. Ideas are pressed away like botanical specimens between boards. "The Bishop," he says, "makes a very telling case against me." "Hurrah! A seed has just germinated after twenty-one and a half hours in an owl's stomach." "I am like a gambler and love a wild experiment."

"I am horribly afraid." "I trust to a sort of instinct and God knows can seldom give any reason for my remarks." And then, pathetically, "All nature is perverse and will not do as I wish it. I wish I had my old barnacles to work at, and nothing new."

The confidences go on like the running stream of the years. Volumes of them exist. There are other volumes that are lost—the one, for example, that might explain what strange compulsion drove Darwin, in at least one recorded instance, from bed at midnight to come downstairs to chatting guests and correct a trivial shift in opinion that could affect no one.

The floating threads of all the ideas, all the thinking, all the nightmares of hours spent in the endless galleries of books, meet and are gathered up at last in the great book of 1859—that book termed the *Origin of Species,* which Darwin to the last maintained was only a hasty abstract.

Do not judge harshly, he importuned no one in particular. Wait for the Big Book, the real book, where everything will be explained and more, so much more, will be adequately interpreted. And if not that book, then the endless regression through ever larger volumes seems to occupy his mind. His odyssey is endless. There is the earthworm, the orchid, the pigeons from the sporting gentry in the gin palace, the shapes of leaves, the carnivorous sundew feeding like an animal. There is the weird groping of vines. There is man himself, the subject of subjects. Year after year he brings his treasures forward like a child and proffers them.

In one or another guise he is hiding from the public.

He has become, save for his family, almost a recluse. To his astonishment, he is a legend in his lifetime. The "abstract" he scorned has become a classic of the world's literature. His name is coupled with that of Newton.

Variation—that subtle, unnoted shifting of the shapes of men and leaves, bird beaks and turtles, that he had pondered over far off in the Circean Galápagos—was now seen to link the seemingly ridiculous and chaotic world of life into a single whole. Selection was the living screen through which all life must pass. No fact could be left a fact; somewhere in the world it was tied to something else. What made a tuft of feathers suddenly appear on a cock's head or induce the meaningless gyrations of a tumbler pigeon? What, in this final world of the fortuitous, the sad eyes questioned, had convinced a tailless ape that he was the object of divine attention?

The year 1859 was gone. The British Association had met at Oxford in 1860 and Thomas Huxley, Darwin's devastatingly caustic defender, had clashed with Bishop Samuel Wilberforce. A lady had fainted. Captain Robert Fitzroy, himself a Fellow of the Royal Society, now a meteorologist and advocate of storm warnings, had arisen angrily to protest the violation of the first chapter of the Book of Genesis.

In a few more years, worn out with his attempts to convince the indifferent Admiralty and public that weather might be foretold and disasters minimized, Fitzroy would go alone upstairs in the dawn with a razor cold against his throat while his family slumbered.

Was he, too, feeling the sickness of that grinding human displacement? Did he, too, feel his solid Victorian world slipping beneath his feet, and, irony of ironies, all because he, Robert Fitzroy, had invited an earnest young man from Cambridge to go on a sea voyage years and years ago? In the dim light the razor glinted. The most important storm warning of all had failed Fitzroy. He was a scientific pioneer who had the misfortune to die unrecognized. He had pleaded for the use of the telegraph in following the weather. Radio and aircraft would prove his wisdom in the following century. Fitzroy had the misfortune to be both behind and ahead of his time. Such men are always subject to injustice.

3

We today know the result of Darwin's endeavors—the knitting together of the vast web of life until it is seen like the legendary tree of Igdrasil, reaching endlessly up through the dead geological strata with living and related branches still glowing in the sun. Bird is no longer bird but can be made to leap magically backward into reptile; man is hidden in the lemur, lemur in tree shrew, tree shrew in reptile; reptile is finally precipitated into fish.

But then there intrudes another problem: Mouse is trying to convert all organic substance into mouse. Black snake is trying to convert mouse into snake. Man maintains factories to convert cattle into human substance. It is an ingenious but hardly edifying spectacle in which nothing really wins, and through which whole orders of life have perished. If our tempo of seeing could be speeded, life would appear and disappear as a chaos of evanescent and writhing forms, possessing the impermanence of the fairy mushroom circles that spring up on our lawns at midnight.

But this is not all; there is something more terrible at the heart of this seething web, something that caused even Darwin's nerves to shudder. Let me illustrate from an experience of my own. I had been far outward on an open prairie swept by wind and sun. The day was fair. It was good to be alive. On all that wild upland I finally saw one human figure far away. He, like myself, seemed wandering and peering. Eventually he approached. We exchanged amenities, and I learned he was a brother scientist, though from a different discipline. He carried dissecting instruments and a bottle.

There was something living in the bottle amidst the juices drained from a rabbit's belly. The man was happy; his day had been successful. He showed me his treasure there in the bright sun on the turf of that sterile upland where there had seemed to lurk no evil. The sight lies a quarter of a century away and still I remember it—the dreadful pulsing object in the bottle and the grinning delight of its possessor.

It was a parasite—a new and unnamed parasite—explained the happy government entomologist, or so the man had represented himself. But there in the bottle, alternately expanding and contracting, searching blindly for the living flesh from which it had been torn, groped an enormous, glistening worm. I was younger then, and my mind clearer. Within it arose the memory of perhaps the most formidable words Darwin was ever to write—words never intended for publication. His personal discovery, in what must have been all its dire immediacy, lay in that moment before me, just as his words re-echoed in my brain. "What a book," he had written with unaccustomed savagery, "a devil's chaplain might write on the clumsy, wasteful, blundering and horribly cruel works of Nature."

Evil has always been one of the difficult questions associated theologically with the fall of man, but now it had been found in the heart of nature itself. It was as though the pulse of the universe had been transferred to that obscene, monstrous body that was swelling as though to engulf the world. The more I looked, the more it appeared to grow. The high clean air of that lonely upland only made the event more unnatural, the collector's professional joy more maniacal.

Finally he went away, bearing his precious bottle with its pulsing, unspeakable life. I watched him until he dwindled to a trudging speck on the distant horizon. It is plain, however, in the light of decades, that that remote, gesticulating figure within my mind is still obscurely laughing in his long descent. If it were not so, I would not see him, for the plain itself was gigantic. This, too, I remember. It would be hard, it seems in retrospect, for anything at all to be found there—just anything, including man. I do not comprehend to this day what it was that ordained our meeting.

Except one thing, perhaps: the fall of man. In that fall, gazing upon the creature in the bottle, I, too, had participated. For man *did* fall; even to an unbeliever and an evolutionist like Darwin. Man fell from the grace of instinct into a confused and troubled cultural realm beyond nature, much as in the old theology man fell from a state of innocence into carnal knowledge. The idea is implicit in Darwin's work and has been commented upon by the critic Stanley Hyman, who has noted its recognition by a late Victorian scholar. This reviewer had commented that "Mr. Darwin [in *The Descent of Man*] finds himself compelled to reintroduce a new doctrine of the fall of man. He shows that the instincts of the higher animals are far nobler than the habits of the savage races of men, and he finds himself, therefore, compelled to reintroduce—in a form whose substantial orthodoxy appears to be quite unconscious—the doctrine that man's gain of knowledge was the cause of the temporary but long enduring moral deterioration . . . of savage tribes." The slow climb back to respectability seems, as one studies Darwin's work, to have culminated in Victorian civilization. The savages of Cape Horn were hardly accorded the graces of domesticated animals.

Darwin had been gazing backward upon the ways of pigeons, apes, and earthworms in extravagant profusion. Upon these creatures and their origins he had expended a sizable proportion of his life. Nature he loved, but by his own words he had become a hermit. Man he achingly endured, as he endured the visitors at Down. He was looking back upon an increasingly remote and violent past, through spectacles few men had raised to their eyes before, and none before him so effectively. Though he occasionally rendered lip service to the idea of human progress toward perfection, cultural man was really a disturbing element in his system, an obstruction difficult to account for, and introducing strange vagaries into Darwin's own version of the Newtonian world machine. In spite of a vast world journey, enormous reading, and a wonderful glimpse, as through the mosaic of a stained-glass window, at the imperfect changing quality of life, Darwin remained an observer held in the bonds of the European social system of his day, and overimpressed by Malthusian struggle. The oncoming world of the indeterminate and the possible that he had helped to initiate he never fully grasped. He looked, and his spectacles brought him light, but it was sometimes the half-light with which Oz has so frequently chosen to shade the eyes of men.

4

Thoreau had loved nature as intensely as Darwin and perhaps more personally. He had seen with another set of glasses. He was, in an opposite sense to Darwin, a dweller along the edge of the known, a place where the new begins. Thoreau carries a hint of that newness. He dwelt, without being quite consciously aware of it, in the age after tomorrow. His friends felt universally baffled by Thoreau and labeled him "almost another species." One contemporary wrote: "His eyes slipped into every tuft of meadow or beach grass and went winding in and out of the thickest undergrowth, like some slim, silent, cunning animal." It has been said that he was not a true naturalist. What was he, then? The account just quoted implies a man similar to Darwin, and, in his own way, as powerfully motivated.

Of all strange and unaccountable things, Thoreau admits his efforts at his *Journal* to be the strangest. Even in youth he is beset by a prescient sadness. The companions who beguile his way will leave him, he already knows, at the first turn of the road. He was basically doomed all his life to be the Scarecrow of Oz, and if he seems harsher than that genial figure, it is because the city he sought was more elusive and he did not have even the Cowardly Lion for company. He knew only that by approaching nature he would be consulting, in every autumn-leaf fall, not alone those who had gone before him, but also those who would come after. He was writing before the *Origin of Species,* but someone had sewn amazing eyes upon the Concord Scarecrow.

There is a delicacy in him that is all his own. His search for support in nature is as diligent as that of a climbing vine he had once watched with fascinated attention groping eerily toward an invisible branch. Yet, like Darwin, he had witnessed the worst that nature could do. On his deathbed he had asked, still insatiable, to be lifted up in order that he could catch through the window a glimpse of one more spring.

In one passage in the *Journal* he had observed that the fishers' nets strung across the transparent river were no more intrusive than a cobweb in the sun. "It is," he notes, "a very slight and refined outrage at most." In their symmetry, he realizes, they are a beautiful memento of man's presence in nature, as wary a discovery as the footprint upon Crusoe's isle. Moreover, this little symbol of the fishers' seine defines precisely that delicately woven fabric of human relationships in which man, as a social animal, is so thoroughly enmeshed. There are times when, intellectually, Darwin threshed about in that same net as though trapped by a bird spider in his own forested Brazil.

For the most part, the untraveled man in Concord managed to slip in and out of similar meshes with comparative ease. Like some lean-bodied fish he is there, he is curiously observant, but he floats, oddly detached and unfrightened, in the great stream. "If we see nature as pausing," Thoreau remarks more than once, "immediately all mortifies and decays." In that nature is man, merely another creature in perspective, if one does not come too close, his civilizations like toadstools springing up by the road. Everything is in the flowing, not the past.

Museums, by contrast, are catacombs, the dead nature of dead men. Thoreau does not struggle so hard as Darwin in his phylogenies to knit the living world together. Unlike the moderns, Thoreau was not constantly seeking nostalgically for men on other planets. He respected the proud solitude of diversity, as when he watched a sparrow hawk amusing itself with aerial acrobatics. "It appeared to have no companion in the universe and to need none but the morning," he remarked, unconsciously characterizing himself. "It was not lonely but it made all the earth lonely beneath it."

Or again, he says plainly, "fox belongs to a different order of things from that which reigns in the village." Fox is alone. That is part of the ultimate secret shared between fox and scarecrow. They are creatures of the woods' edge. One of Thoreau's peculiar insights lies in his recognition of the creative loneliness of the individual, the struggle of man the evolved animal to live "a supernatural life." In a sense, it is a symbolic expression of the equally creative but microcosmic loneliness of the mutative gene. "Some," he remarks, "record only what has happened to them; but others how *they* have happened to the universe."

To this latter record Thoreau devoted the *Journal* that mystified his friends. Though, like Darwin, he was a seeker who never totally found what he sought, he had found a road, though no one appeared to be walking in it. Nevertheless, he seems to have been interiorly informed that it was a way traversed at long intervals by great minds. Thoreau, the physical stay-at-home, was an avid searcher of travel literature, but he was not a traveler in the body. Indeed, there are times when he seems to have regarded that labyrinth—for so he called it—with some of the same feeling he held toward a house—as a place to escape from. The nature we profess to know never completely contained him.

"I am sensible of a certain doubleness," he wrote, "by which I can stand as remote from myself as from another. . . . When the play—it may be the tragedy of life—is over, the spectator goes his way. It was a kind of fiction." This man does suggest another species, perhaps those cool, removed men of a far, oncoming century who can both live their lives and order them like great art. The gift is rare now, and not wholly enticing to earthbound creatures like ourselves.

Once, while surveying, Thoreau had encountered an unusual echo. After days with humdrum companions, he recorded with surprise and pleasure this generosity in nature. He wanted to linger and call all day to the air, to some voice akin to his own. There needs must be some actual doubleness like this in nature, he reiterates, "for if the voices which we commonly hear were all that we ever heard, what then? Echoes . . . are the only kindred voices that I hear."

5

Here, in Thoreau's question, is the crux, the sum total of the human predicament. This is why I spoke of our figuratively winding our way backward into a spiritual winter, why I quoted an Eskimo upon wisdom. On the eve of the publication of the *Origin of Species,* Thoreau, not by any means inimical to the evolutionary philosophy, had commented: "It is ebb tide with the scientific reports."

In some quarters this has aroused amusement. But what did Thoreau mean? Did he sense amidst English utilitarian philosophy, of which some aspects of Darwinism are an offshoot, an oncoming cold, a muffling of snow, an inability to hear echoes? Paradoxically, Thoreau, who delighted in simplicity of living, was averse to the parsimonious nature of Victorian science. It offended his transcendental vision of man. Lest I seem to exaggerate this conflict, read what Darwin himself admitted of his work in later years:

"I did not formerly consider sufficiently the existence of structures," he confesses, "which as far as we can . . . judge, are neither beneficial nor injurious, and this I believe to be one of the greatest oversights as yet detected in my work. This led to my tacit assumption that every detail of structure was of some special though unrecognized service."

We know that Thoreau already feared that man was becoming the tool of his tools, which can, alas, include ideas. Even now, forgetting Darwin's belated caution, those with the backward-reaching spectacles tell us eagerly, if not arrogantly, in the name of evolution, how we are born to behave and the limitations placed upon us—we who have come the far way from a wood nest in a Paleocene forest. Figuratively, these pronouncements have about them the enlarging, man-destroying evil of the pulsing worm. They stop man at an imagined border of himself. Man suffers, in truth, from a magical worm genuinely enlarged by a certain color of spectacles. It is a part, but not the whole, of the magic of Oz.

Is it not significant, in contrast to certain of these modern prophets, that Thoreau spoke of the freedom he felt to go and come in nature; that what is peculiar to the life of man "consists not in his obedience, but his opposition to his instincts"? The very behavior of the other animals toward mankind, Thoreau knew, revealed that man was not yet the civilized creature he pretended to be.

One must summarize the two philosophies of evolution and then let the Eskimo speak once more. In the Viking Eddas it is written:

> Hard it is on earth . . .
> Ax-time, sword-time . . .
> Wind-time, wolf-time, ere the world falls
> Nor ever shall men each other spare.

Through these lines comes the howl of the world-devouring Fenris-wolf, waiting his moment under the deep-buried rocket silos of today. In the last pages of *Walden* one of Thoreau's wisest remarks is upon the demand scientific intellectuals sometimes make, that one must speak so as to be always understood. "Neither men," he says, "nor toadstools grow so." There is a constant emergent novelty in nature that does not lie totally behind us, or we would not be what we are.

Here is where Thoreau's sensitivity to echoes emerges powerfully: It is onflowing man, not past evolutionary man, who concerns him. He wants desperately to know to what degree the human mind is capable of inward expansion. "If the condition of things which we were made for is not yet at hand," he questions anxiously, "what can we substitute?" The echoes he senses are reverberating from the future.

Finally, he compresses into a single passage the answer to the wolf-time philosophy, whether expressed by the Viking freebooters, or by certain of their modern descendants. "After," he says, "the germs of virtue have thus been prevented many times from developing themselves, the beneficent breath of evening does not suffice to preserve them. . . . Then the nature of man does not differ much from that of the brute."

Does this last constriction contain the true and natural condition of man? No, Thoreau would contend, for nature lives always in anticipation. Thoreau was part of the future. He walked toward it, knowing also that in the case of man it must emerge from within by means of his own creation. That was why Thoreau saw the double nature of the tool and eyed it with doubt.

The soul of the universe, the Upholder, reported Rasmussen of the Alaskan Eskimo, is never seen. Its voice, however, may be heard on occasion, through innocent children. Or in storms. Or in sunshine. Both Darwin and Thoreau had disavowed the traditional paradise, and it has been said of Thoreau that he awaited a Visitor who never came. Nevertheless, he had felt the weight of an unseen power. What it whispers, said the men of the high cold, is, "Be not afraid of the universe."

Man, since the beginning of his symbol-making mind, has sought to read the map of that same universe. Do not believe those serious-minded men who tell us that writing began with economics and the ordering of jars of oil. Man is, in reality, an oracular animal. Bereft of instinct, he must search constantly for meanings.

We forget that, like a child, man was a reader before he became a writer, a reader of what Coleridge once called the mighty alphabet of the universe. Long ago, our forerunners knew, as the Eskimo still know, that there is an instruction hidden in the storm or dancing in auroral fires. The future can be invoked by the pictures impressed on a cave wall or in the cracks interpreted by a shaman on the incinerated shoulder blade of a hare. The very flight of birds is a writing to be read. Thoreau strove for its interpretation on his pond, as Darwin, in his way, sought equally to read the message written in the beaks of Galápagos finches.

But the messages, like all the messages in the universe, are elusive.

Some months ago, walking along the shore of a desolate island off the Gulf Coast, I caught a glimpse of a beautiful shell, imprinted with what appeared to be strange writing, rolling in the breakers. Impelled by curiosity, I leaped into the surf and salvaged it. Golden characters like Chinese hieroglyphs ran in symmetrical lines around the cone of the shell. I lifted it up with the utmost excitement, as though a message had come to me from the green depths of the sea.

Later I unwrapped the shell before a dealer in antiquities in the back streets of a seaport town.

"*Conus spurius atlanticus,*" he diagnosed for me with brisk efficiency, "otherwise known as the alphabet shell."

But why spurious? I questioned inwardly as I left the grubby little shop, warily refusing an offer for my treasure. The shell, I was sure, contained a message.

We *live* by messages—all true scientists, all lovers of the arts, indeed, all true men of any stamp. Some of the messages cannot be read, but man will always try. He hungers for messages, and when he ceases to seek and interpret them he will be no longer man.

The little cone lies now upon my desk, and I handle it as reverently as I would the tablets of a lost civilization. It transmits tidings as real as the increasingly far echoes heard by Thoreau in his last years.

Perhaps I would never have stumbled into so complete a revelation save that the shell was *Conus spurius,* carrying the appellation given it by one who had misread, most painfully misread, a true message from the universe. Each man deciphers from the ancient alphabets of nature only those secrets that his own deeps possess the power to endow with meaning. It had been so with Darwin and Thoreau. The golden alphabet, in whatever shape it chooses to reveal itself, is never spurious. From its inscrutable lettering is created man and all the streaming cloudland of his dreams.

SEVEN

The Invisible Island

I say by Sorcery he got this Isle;
From me he got it.

—CALIBAN

IN latitude sixty degrees south, under the looming shadow of the antarctic ice, from somewhere deep in the freezing planetary swirl of the Humboldt Current, the great whale surfaced and was struck for the last time. "We found embedded in her hide," the record of the old logbook runs, "divers harpoons of antique shapes, ours being her death."

What shafts have equally been hurled in epochs gone against the quarry earth? What gaping prehistoric jaws, subsiding in oblivion, have struck and splintered in their turn upon her stone? What giant plinths and cromlechs buried in her soil still mark the deeds of conquerors long since undone? What mile-deep drills and missiles wrought by hands more violent now probe at earth's defenses? Yet she flies on regardless. Her hour has not yet come. The blade that will split her heart has not been forged.

Earth is the mightiest of the creatures. She contains beneath her furry hide the dark heart of nothingness, from which springs all that lives. She is the wariest and most complete of animals, for she lends herself to no particular form and in the end she soundlessly forsakes them all. She is the one complete island of being. The rest, including man, are in some degree fragmented and illusory. Yet it is these phantoms, even of whales come up from the deep gulfs, that mark our thought.

Of true places Herman Melville once remarked that "they are never down on any map." He also observed of those great amphibians, those islands of living flesh of which I have just spoken, that it is not known whether the small eyes placed yards apart on opposite sides of their huge heads can co-ordinate the meaning of two sights at once. If not, that great solitary beast, with all its hoary antiquity written in the weapons ranged along its back, saw, as it floundered in its death throes, two separate worlds. It saw with one dimming affectionate eye the ancient mother, the heaving expanse of the universal sea. With the other it glimpsed with indescribable foreboding the approaching shape of man, the messenger of death and change. So it was that the dying whale in its dissociated vision had arrived, if momently, at the one true place where the nature of the past mingles with the onrush of the future and is borne down forever into darkness. It is an instant to remember, for it will come, in turn, to man.

Indeed, this fusing of the past and future has already come and made of man the thing he is, an invisible island, as surely as the great whale was an island, as surely as volcanic clinkered isles produce monstrosities, dwarfs, and giants in secret shifting latitudes where no navigator is able to take readings. It is a somber reflection upon human nature that so much has been written about the triumph of the fittest and so little about the survival of the failures who have changed, if not deranged, the world.

Man is one such creature, but the story begins long before him, a story of only incidental triumphs. The universe, as we have seen, is a place of unexpected events. A major portion of the world's story appears to be that of fumbling little creatures of seemingly no great potential, falling, like the helpless little girl Alice, down a rabbit hole or an unexpected crevice into some new and topsy-turvy realm. Such beings sometimes find themselves cast unhappily in the role of unanticipated giants. It is not struggle and survival alone that have so marked life's exuberant pathway; rather, the mastery has often come after the event and almost as a prelude to extinction. It is as though everything alive had in it a tug of antigravity, a revulsion from the central fire or the mother sea. If stars and galaxies hurtle outward in headlong flight, the urge for dispersal seems equally and unexplainably written into the living substance itself. The myth of Eden, the myth of Babel, and now our evolving science itself all testify to this rent in nature, as though, if we could but see it, everything might be scurrying and hopping toward a myriad exits. Solitude may strike self-conscious man as an affliction, but his march is away from his origins, and even his art is increasingly abstract and self-centered. From the solitude of the wood he has passed to the more dreadful solitude of the heart.

In a university lecture hall some months ago, it was brought home to me just how far this alienation had progressed. I had been speaking, by way of illustrating a point, about a tiny deer mouse, a wonderfully new and radiant little creature of white feet and investigative fervor whom I had seen come into a basement seminar upon the Byzantine Empire. After a time the mouse in its innocent pleasure had actually ascended into an empty chair and perched upright with trembling inquisitive gravity while an internationally renowned historian continued to address the group. By no least sign did he reveal that an eager anticipatory face had appeared among his students.

After my own lecture I was approached and chided by a young lady who informed me with severity that I was betraying evidence of a foolish anthropomorphism, which would certainly place me under disfavor and suspicion in the psychological circles she frequented.

I sighed and reluctantly confessed that perhaps the mouse, since he was obviously a very young mouse recently come from the country, could not have understood every word of the entire lecture. Nevertheless, it was gratifyingly evident to my weary colleague, the great historian, that the mouse had at least tried.

"You see," I explained carefully, "we may have witnessed something like Alice in reverse. The mouse came through a crevice in the wall, a chink in nature. Man in his time has come more than once through similar chinks. I admit that the creatures do not always work out and that the chances seemed rather against this one, but who is to say what may happen when a mouse gets a taste for Byzantium rather beyond that of the average graduate student? It takes time, generations even, for this kind of event to mature.

"If I may be pardoned for being so bold," I remonstrated with the young lady, "what do you think your chances might have been of charging *me* with anthropomorphism when we were both floundering about in a mud puddle, or, for that matter, testing whether an incipient backbone might enable us to wriggle upstream? You must remember," I continued, "that these are all figurative entrances and exits with sometimes kingdoms at the bottom of them. Or disaster, or even both together."

"But not," the young lady protested venomously, "the Byzantine Empire."

"My friend was quite hopeful," I persisted. "The creature was so evidently concerned and alert beyond the average. After all, how must the first two invented words have stirred your ancestors," I appealed, "and there was nobody, absolutely nobody, to give them a lecture on Byzantium, because, well, because Byzantium was"—I was getting out of my depth now—"up there in the future." I gestured uncertainly with one finger.

"It was not down a hole then," said the literal young

lady. Triumphantly she drove home her point. "It wasn't anywhere." With this she walked out.

"This woman is evidently part of a conspiracy to keep things just as they are," I later wrote to my friend. "This is what biologically we may call the living screen, the net that keeps things firmly in place, a place called now.

"It doesn't always work," I added in encouragement. "Things get through. We ourselves are an example. Perhaps a bad one. About the mouse . . ."

The answer came back in a few days, lugubriously. "The Exterminators have come. Your chink is closed. Definitely."

The Exterminators. I turned the harsh word over in my mind. Great God, what were we doing? The net was drawing tighter. Man had his hands upon it. The effects were terrifying. I thought of a sparkling stream where I had played as a boy. There had been sunfish in it, turtles. Now it ran sludge and oil. It was true. The net was tightening. All over the earth it was tightening. Even the ice chinks at the poles were under man's surveillance.

One might say that the surveillance began in 1835, when young Charles Darwin, on the round-the-world cruise of H.M.S. *Beagle,* dropped anchor in the Galápagos Archipelago, six hundred miles off the west coast of South America. Ashore, young Darwin observed with a speculative eye a bird strange to his experience of the continents. He jotted into his notebook an observation that was not to bear full fruit until after the passage of over a quarter of a century.

Such island differences among the creatures of the Archipelago, the young man meditated, "*will be well worth examining; for such facts would undermine the stability of species.*"

2

I have often looked with speculative interest upon those delicate insects that row with feathery feet upon the waters of a brook. They make but a slight dimple upon the film of sliding water. They breathe the air; they rove in an immense freedom over ponds and watercourses. The insubstantial film upon which they float resembles the surface tension of the living screen of life, in which every organism, like the forces in the atom, exerts an enormous hold, directly or indirectly, upon every other living thing. The water striders have evolved a way of outwitting the water film that entraps the heavy-footed. In their small way they have risen superior to a dangerous medium and have diverted its tensions to their own advantage.

Man has similarly defeated and diverted the entire web of life and dances, dimpling, over it. Like the water strider he possesses the freedom of a dangerous element. Even the water strider's freedom is relative, however, and contained within nature. One cannot help but dwell upon the hidden powers that produce so delicate a balance between freedom and catastrophe. For freedom of this nature is rare, and in man it is more than rare—it is unique—for he dances upon shadows, the shadows in his brain.

Before turning to that realm of shadows, it is well to define what we mean by the web of life. Some time ago I had occasion one summer morning to visit a friend's grave in a country cemetery. The event made a profound impression upon me. By some trick of midnight circumstance a multitude of graves in the untended grass were covered and interwoven together in a shimmering sheet of gossamer, whose threads ran indiscriminately over sunken grave mounds and headstones.

It was as if the dead were still linked as in life, as if that frail network, touched by the morning sun, had momentarily succeeded in bringing the inhabitants of the grave into some kind of persisting relationship with the living. The night-working spiders had produced a visual facsimile of that intricate web in which past life is intertwined with all that lives and in which the living constitute a subtle, though not totally inescapable, barrier to any newly emergent creature that might attempt to break out of the enveloping strands of the existing world.

As I watched, a gold-winged glittering fly, which had been resting below the net, essayed to rise into the sun. Its wings were promptly entrapped. The analogy was complete. Life *did* bear a relationship to the past and was held in the grasp of the present. The dead in the grass were, figuratively, the sustaining base that controlled the direction of the forces exerted throughout the living web.

The golden fly would perish. Nevertheless, there would be days when the wind would sweep the net away, or when snow would hold the sleeping cemetery in a different, though even more profound, embrace. Everything, in the words of John Muir, tended to be hitched to everything else. Yet this was not all, or life would not have originally engaged itself upon that exploratory adventure we call evolution. The strands of the living web were real; they did check, in degree, the riotous extremes of variation. Fortunately for the advancement of life, the tight-drawn strands sometimes snapped; the archaic reptile gave place to the warm-blooded mammal.

Sometimes it was less the snapping of a thread, even of gossamer, than the secretive discovery of a hidden doorway, found by some blundering creature pushed to the wall by savage competitors in the seemingly impermeable web. An old-fashioned, bog-trapped fish had managed to totter ashore on its fins and find itself safely alone to develop in a new element. A clumsy, leaping reptile had later managed to flounder into the safety of the air.

This is the fascinating rabbit-hole aspect of the living world. Most of its experiments are small, at best localized adjustments that, if anything, draw the mesh of the living screen by degrees ever tighter. Man, once having successfully escaped the net, has busied himself in drawing the meshes together with a strangling intensity exerted upon the rest of life. He now appears to possess the power to tear the net at his pleasure. Yet as I stood in wonder before the entangled tombstone of my friend, it struck me that not even man can escape completely the Laocoön embrace of the living web. Sir Francis Bacon in the early dawn of experimental science had called attention to the fact that nature was capable on occasion of extreme "exuberances," such as would be well exemplified by human expansion today. It did not escape the Elizabethan philosopher that such irregularities are by the nature of things eventually, in their turn, trapped and confined.

Life as we know it is not limitless in its capacities, above all, in the higher organisms. Actually, its manifestations are confined to a small range on the thermometer. Moreover, life demands water for the maintenance of its interior environment, in quantities hard to come by, if this solar system can be taken as a typical example. Long food chains, from microscopic infusoria to whales, sustain and nourish all manner of strange animals and plants. Man himself is not free from these food chains, though increasingly, and sometimes to his peril, he tampers experimentally with his environment —thus inviting that "violence in the Returne" of which Bacon warned.

Life has survived by distributing itself over innumerable tiny environments. It has mastered such extremes of temperature and pressure as are represented by the abyssal depths of the sea, by deserts where water must be hoarded behind impermeable plant walls, or where breath must be drawn painfully upon Himalayan heights. In these circumstances one fact is self-evident: under such trials life tends to thin out to the point of disappearance.

In that disappearance the forms of the highest nervous complexity are the first to vanish. All mutative changes tend to be directed toward holding the unfavorable outside world at bay and preserving by formidable barriers the life within. The almost complete rejection of that outside world, however—no matter how cleverly the organism has adapted its resources for the purpose of sustaining itself under such inhospitable conditions—spells the end of intense conscious interaction with the environment. One is left with drought-resisting plants, infusoria that can survive desiccation, crabs teetering gingerly over nightmare depths of mud, or a few desert burrowers capable of synthesizing minute quantities of water in their tissues. Still, the tenuous lines of the web of life run up into mountainous heights and descend into the depths of the sea.

The living screen known as the biosphere may best be pictured by dipping some rough-skinned fruit, such as an orange, into water. Upon its withdrawal the wet film adhering to the depressions of the orange gives some idea of the relative thickness of the living membrane that covers most of the planet. It will attenuate toward the poles and upon ice-covered peaks; it will slink underground or disappear in the most extreme and shifting deserts.

It is a tenuous film, nothing more—yet a film of incredible complexity. Life survives through the fact that we possess an oxygen-charged atmosphere, a modest temperature, and the seas so beautifully photographed by the homing astronauts. We rotate in the beneficent light of a minor star. On freezing Pluto, our outermost planet, that star would appear about the size of a nailhead. Even in the auspicious environment of earth, the major portion of the ancestral past is represented only in eroding strata. The net, the living film that at any given moment of earth time seems to hold all the verdant world in an everlasting balance, is secretly woven anew from age to age. The tiny spiders that had worked through a single night amidst the lettered stones of the cemetery had reproduced in miniature the tight-strung gossamer that links us with the remains of our animal past.

We have noted that when Darwin intruded into the Galápagos and observed the biological rarities existing there, his growing suspicion about the reality of evolution was confirmed. By the time he wrote the *Origin of Species* and *The Descent of Man,* his attention had been occupied by his theory of how evolution had come about; namely, through natural selection. Natural selection was defined as the preservation of favorable genetic variations by means of the winnowing effect of the struggle for existence.

We need not here pursue the intricacies of modern genetic theory. In Darwin's time the emphasis upon intragroup struggle as a means of evolutionary advance sometimes took on exaggerated forms. It has tended so to re-emerge in simplistic interpretations of modern human problems. In the nineteenth century Moritz Wagner reminded Darwin of the bearing that isolation must exercise upon biological change. Darwin, in turn, responded, "it would have been a strange fact if I had overlooked the importance of isolation seeing that it was such cases as that of the Galápagos Archipelago, which chiefly led me to study the origin of species."

Nevertheless, by his own confession, he had "oscillated much" between what seemed to be the intense struggle present on the continents compared with the relaxed refuge area of the Galápagos. In reality, natural selection is a broad phrase, so broad that it implies many different kinds of change, or even repression of change, as when no opening exists in the constricting web we have examined. Islands are apt by their seclusion to offer doorways to the unexpected, rents in the living web, opportunities presented to stragglers who might be carrying concealed genetic novelty in their bodies—novelty that might have remained suppressed in a more drastic competitive environment.

Competition may simply suppress what exists only as potential. The first land-walking fish was, by modern standards, an ungainly and inefficient vertebrate. Figuratively, he was a water failure who had managed to climb ashore on a continent where no vertebrates existed. In a time of crisis he had escaped his enemies. For a long time he was in a position to perfect freely responses to his new mode of life. Adaptive radiation and struggle would only emerge later, as our fishy forerunner explored other land environments and began to create, in his turn, a new repressive web.

To have a genetic island there must be in the beginning an isolating barrier. It could be a genuine island, such as has occupied the speculative attention of voyagers since the days of Captain Cook. On the other hand, the "island" may be a mountaintop or the product of a glacial obstruction. An impassable stream may suffice, or the barrier of a season. The boundary may be also singularly present but invisible. Whatever else it may be, it must offer the opportunity of creative genetic change on a large scale if life is to advance. Struggle, of and by itself, does little but sharpen what exists to a superior efficiency. True, it plays an important role in evolution, but it is not necessarily the only, or even the primary, factor in the rare emergence of the completely novel. It must always be remembered that natural selection is one of those convenient magical phrases that can embrace both dramatic change and stultifying biological conservatism.

This morning, on my back lawn, giant mushrooms of a species unfamiliar have pushed up through the grass and leaves. They have the appearance of distorted livers, or of some other unsightly organ. A single fog-filled night has sufficed to produce them. If the weather lasts, there will be more. Probably their spores have been merely waiting—waiting for who knows how long. The visitation of the fog represents a sudden rent in the living screen, an opportunity that fungi, as creatures of the night, are particularly suited to seize. The fact that the season is close onto winter makes their sudden upheaval intensely dramatic. It makes one ask what kind of word spores, what night-fog in the protohuman cranium, induced the emergence of that fantastic neurofungus which is man.

3

Islands can be regarded as something thrust up into recent time out of a primordial past. In a sense, they belong to different times: a crab time or a turtle time, or even a lemur time, as on Madagascar. It is possible to conceive an island that could contain a future time —something not quite in simultaneous relationship with the rest of the world. Perhaps in some obscure way everything living is on a different time plane. As for man, he is the most curious of all; he fits no plane, no visible island. He is bounded by no shore, except a shore of shadows. He has emerged almost as soundlessly as a mushroom in the night.

Islands are also places of extremes. They frequently produce opposites. On them may exist dwarf creations induced by lack of space or food. On the other hand, an open ecological niche, a lack of enemies, and some equally unopposed genetic drift may as readily produce giantism. The celebrated example of the monster Galápagos turtles comes immediately to mind. Man constitutes an even more unique spectacle for, beginning dwarfed and helpless within nature, he has become a Brocken specter as vast, murk-filled, and threatening as that described in old Germanic folk tales. Thomas De Quincey used to maintain that if one crossed oneself, this looming apparition of light and mist would do the same, but with reluctance and, sometimes, with an air of evasion. One may account this as natural in the case of an illusion moving against a cloud bank, but there is, in its delayed, uncertain gesture, a hint of ambiguity and terror projected from the original human climber on the mountain. To the discerning eye there is, thus, about both this creature and his reflection, something partaking of both microscopic and gigantic dimensions.

In some such way man arose upon an island—not on a visible oceanic island but in some hidden forest meadow. Man's selfhood, his future reality, was produced within the invisible island of his brain—the island clouded in a mist of sound. In this way the net of life was once more wrenched aside so that an impalpable shadow quickly wriggled through its strands into a new, unheard-of dimension of existence. Following this incredible event the natural world subsided once more into its place.

There was, of course, somewhere in the depths of time, a physical location where this episode took place. Unlike the sea barriers that Darwin had found constricting his island novelties, there had now appeared a single island whose shores seemed potentially limitless. This island, no matter where it had physically arisen, had been created by sound vibrating with meaning in the empty air. The island was based on man's most tremendous tool—the word. By degrees, the word would separate past from present, project the unseen future, contain the absent along with the real, and define them to human advantage. Man was no longer confined, like the animal, to what lay before his eyes or his own immediate attention. He could juxtapose, divide, and rearrange his world mentally. Upon the wilderness of the real, men came to project a phantom domain, the world of culture. In the end, their cities would lie congregated and gleaming like the nerve ganglions of an expanding brain.

Words would eventually raise specters so vast that man cowered and whimpered, where, as an animal, he had seen nothing either to reverence or to fear. Words expressed in substance would widen his powers; words, because he used them ill, would occasionally torture and imprison him. They would also lift him into regions of great light. Even the barbarian north, far from the white cities of the Mediterranean, would come to speak in wonder of the skald who could unlock the "word hoard." In our day this phantom island has embraced the planet, a world from which man has begun to eye the farther stars.

Isolation has produced man just as, on the Atlantic island of South Trinidad, there has emerged, as recorded by the late Apsley Cherry-Garrard, another and most eerie world, a world completely usurped by land crabs. Here, instead of a universe dominated by huge but harmless tortoises, the explorer encountered a nightmare. "The crabs," he recounted, "peep out at you from every nook and boulder. Their dead staring eyes follow your every step as if to say, 'If only you will drop down we will do the rest.' To lie down and sleep on any part of the island would be suicidal. . . . No matter how many are in sight they are all looking at you, and they follow step by step with a sickening deliberation."

Such remarkably confined but distinct worlds upon a single planet are infinitely informative. Because they are geologically younger than the planet itself, they tell us that life's power to create the new is still smoldering unseen about us in the living present. Darwin saw in the Galápagos Islands a reptilian world that had drowsed its antique way into the present. In a sense, Darwin was correct, but the islands are more than this. They are, in miniature, an alternative to the world we know, just as the nightmare crabs upon South Trinidad present an unthinkable alternative to ourselves. Each of these island worlds has taken through chance a different turning at some point in the past; each possesses an emergent novelty. In none has man twice appeared save as a wandering intruder from outside. Man belongs to his own island in nature, the invisible one of a growing genetic isolation about whose origins we know little.

Biological time never creates the same world twice, but out of its clefts and fissures creep, at long intervals, surprisingly original creatures whose destinies can never be anticipated before they arrive. It was so when the first Crossopterygian lungfish inched painfully into the new medium of the air. It was equally true when the first pre-ice-age man-apes drew an abstract line in speech between today and tomorrow. Out of just so much they would succeed in constructing a world.

Impelled by the rising flame of consciousness, they would ingest more than simple food. They would, instead, feed upon the contingent and the possible within their minds. They would step beyond the nature from which they had arisen and eventually turn upon her the level objective gaze of a stranger. The rift so created is widening in our time. It is as though our self-created island were adrift on lawless currents like that same Galápagos Archipelago which seemed, to the old navigators, to set all their calculations at nought.

Over a century ago Thoreau, who had a sensitive ear for the tread of any overgrown Brocken specter in the shape of man, admitted that he would gladly fall "into some crevice along with leaves and acorns." The percipient philosopher felt the need for a renewed hermitage, a natural spot for hibernation. He was seeking a way back through the leafy curtain that has swung behind us, never to open again. Man had broken through that network of strangling vines by the magical utterance of the first word. In that guttural achievement he had created his destiny and taken leave of his kindred. There would be no way of return save perhaps one: through the power of imaginative insight, which has been manifested among a few great naturalists.

4

Much has been spoken of oceanic islands as providing the haven, the sheltering crevice, the bleak and empty quarter in which potentially new forms of life might find the refuge necessary for immediate survival. In the words of one student of island faunas, Sherwin

Carlquist, these creatures, far from being the ripe products of the war of nature, frequently represent, instead, the "inefficient, the unafraid and the obsolescent." How, then, can they be said to have any bearing upon the struggle for existence on the continents—the world where even the astute and perceptive Darwin had difficulty in explaining how the naked and physically inefficient protohumans had survived the claws of the great carnivores?

So puzzled had Darwin found himself that he had hesitated as to whether ancestral man might have been originally a genuine island product. Torn between viewing man as a weak creature needing protection in his early beginnings, or, on the other hand, attempting to visualize him as a fanged gorilloid monster fully equipped for the war of nature on the African land mass, the great biologist had failed to quite perceive the outlines of that invisibly expanding universe which man had unconsciously created out of airy nothing.

As we have seen, it is the wet fish gasping in the harsh air on the shore, the warm-blooded mammal roving unchecked through the torpor of the reptilian night, the lizard-bird launching into a moment of ill-aimed flight that shatter all purely competitive assumptions. These singular events reveal escapes through the living screen, penetrated, one would have to say in retrospect, by the "overspecialized" and the seemingly "inefficient," the creatures driven to the wall. Only after their triumphant planetary radiation is something new observed to have arisen in solitude and silence.

We begin in infancy with a universe that our minds

constantly strive to subdue to the rational. But just as we seem to have achieved that triumph, some part of observed nature persists in breaking out once more into the unexpected. No greater surprise could have been anticipated than protoman's first stumbling venture into the hitherto unglimpsed ghostland of shifting symbols, that puzzling realm which Darwin vainly sought upon some real island in a more substantial sea. Reality has a way of hiding even from its most gifted observers.

In the Hall of Early Man in one of America's great archaeological museums there stands a reconstruction in the flesh of one of the early man-apes of preglacial times. He is reproduced as the bones tell us he must have basically appeared, small of brain, with just one foot, it could be said, across the human threshold. The whole appearance of the creature seems shrunken by modern standards—made more so by a fragment of bone clutched in a small uncertain hand. There remains with the observer an impression of fumbling weakness, as though the pygmy's dreams had depleted him—as though his genuine being had somehow been projected forward across millennia into the Herculean Grecian figure standing beyond him.

The dwarfed man-ape stands on the border of his invisible island; nor can there be any doubt of the nature of that island. It is Prospero's realm, whose first owner was Caliban. It is Shakespeare's island of sweet sounds and miraculous voices. It is the beginning rent in the curtain, the kingdom once confined to a single thicket, the isle of which it was once said, "this is no mortal business."

Some months ago, engaged in travel, I lay in a troubled sleep in the solitary freezing bed of a Canadian hotel. A blizzard had been raging. Beyond the raw new town stretched a stand of dark spruce forest such as one no longer sees in daylight. Somewhere toward dawn I dreamed.

The dream was of a great blurred bearlike shape emerging from the snow against the window. It pounded on the glass and beckoned importunately toward the forest. I caught the urgency of a message as uncouth and indecipherable as the shape of its huge bearer in the snow. In the immense terror of my dream I struggled against the import of that message as I struggled also to resist the impatient pounding of the frost-enveloped beast at the window.

Suddenly I lifted the telephone beside my bed, and through the receiver came a message as cryptic as the message from the snow, but far more miraculous in origin. For I knew intuitively, in the still snowfall of my dream, that the voice I heard, a long way off, was my own voice in childhood. Pure and sweet, incredibly refined and beautiful beyond the things of earth, yet somehow inexorable and not to be stayed, the voice was already terminating its message. "I am sorry to have troubled you," the clear faint syllables of the child persisted. They seemed to come across a thinning wire that lengthened far away into the years of my past. "I am sorry, I am sorry to have troubled you at all." The voice faded before I could speak. I was awake now, trembling in the cold. There was nothing to be seen at the window but the rising flurries of the snow.

Finally they ceased, and all around the little village, wrapped in an enormous eldritch winter, slumbered the dark forest. I looked below my window, but there were no tracks. I looked beside my bed, but there was, in reality, no phone. I lay back, huddled under the blanket, and thought briefly of the shrunken figure of the ape in the museum and of what had once been projected out of his living substance. My own body, in the freezing cold, felt wracked by a similar psychic effort that still persisted. Far off, as over an immense distance in my brain, I heard the echo of my own true voice, or perhaps it was mankind's accumulated voice, for the last time. It was hauntingly beautiful, but it was going. It would not be supplicated. As though in response to my thought, the incessant march of snow began again in the forest.

So this is the end, I thought, wrapping my shoulders closer against the increasing cold; we are, in truth, a failure. Beautiful and terrible, perhaps, but a failure, an island failure—an island whose origins are lost in time yet are still about us in such a way that we do not see them. To an invisible island we owe both triumph and disaster. At first, the island was a rent, a very small rent, in the curtain of life—perhaps no more than a few hairy creatures in a forest glade making experimental sounds that could be varied, one sound that defined the past and another that signified tomorrow. They were very small sounds with which to create an island, but the island, like those fostered by volcanoes on the sea floor, grew, and remained at the same time invisible. The sounds—passing from brain to brain, defining, measuring, remembering how stones could be broken—were responsible. If the island was invisible it was, unlike all other islands, shoreless as it grew. Nevertheless, man, like the other creatures on true islands, was isolated. He drifted insensibly from the heart of things. At first, he kept an uncertain memory of his origins in the animal world. He claimed descent from Grandfather Bear or Raven and on ritual occasions he talked to them or to his mother, the Earth.

The precarious thread that bound man to the living whole finally snapped. He had passed irrevocably into another dimension. His predicament is recognized in the myths of the Tree and of Pandora's box. He had learned to distinguish good from evil. Moreover, his capacity for evil increased as he discovered that the tiny sounds could be made to lie. This was an island within an island. It separated people into many islands. As man entered upon a wild new corridor of existence, some part of himself passed into a hypnotic slumber, but, in the diverse rooms of the mind, other sleepers awakened.

Man has indeed become a giant, but within him, growing at the pace of his own island, is locked the original minuscule dwarf who had stumbled out of the strangling grasp of the forest with a stone clutched in its hand. "This thing of Darknesse," speaks Shakespeare in the shape of the learned Prospero, "I acknowledge mine."

"There is no loneliness," once maintained the Egyptologist John Wilson, "like the loneliness of a mighty place fallen out of its proper service to man." Perhaps the same loneliness inevitably haunts modern man himself, that restless and vacant-eyed wanderer through the streets of cities, that man of ruinous countenance from whom the gods have hidden themselves. Above him, somewhere in the blue, like a hawk hovering over a deserted temple as if in expectation of some divine renewal, the spirit Ariel, dismissed and masterless, still lingers, soaring. Ariel had been long entrapped in the knotted pine that one suspects was human flesh. Does he now hesitate to leave his prison, grown used at last to the drunken mortal cry, "The sound is going. Hark, let's follow it"?

Perhaps it was just such music that I had heard traversing a phantom wire at midnight. Perhaps its purpose was to lead me to another doorway, another opening portal in the dark forest that is man. It is certain that the wooded shores that now confine us lie solely within ourselves. But they are the shores still frequented at midnight by a vengeful Caliban.

EIGHT

The Inner Galaxy

There is strong archaeological evidence to show that with the birth of human consciousness there was born, like a twin, the impulse to transcend it.

—ALAN MCGLASHAN

MANY years ago, I, with another youth of my own age whom I had persuaded to make the journey with me, walked throughout the day up a great mountain. There was a famous astronomical observatory upon the mountain. On certain nights, according to the guide-books, the lay public might come to the observatory and look upon some remote planetary object. They could also hear a lecture.

The youth and I, who had much eager interest but no money, were unable to join one of the numerous tours organized from the tourist hotels in the valley. Instead, we had trudged for many hours in order to arrive before the crowds of visitors might frustrate our hopes for a glimpse of those far worlds about which we had read so avidly.

This was long ago, and we were naïve young men. We thought, though we were poor, that we would be welcome upon the mountain because of our desire to learn. There were reputed to dwell in the observatory men of wisdom who we hoped would receive us kindly since we, too, wished to gaze upon the wonders of outer space. We were, indeed, very unskilled in the ways of the adult world. As it turned out, we were never permitted to see the men of wisdom, or to gaze through the magic glass into outer space. I rather suspect that the eminent astronomers had not taken youths like us into their calculations. There was, it seemed, a relationship we had never suspected between the hotels in the valley and the men who inhabited the observatory upon the mountain.

Although by laborious effort we had succeeded in arriving before the busloads of tourists from the hotels, we were thrust forth and told to take our chances after the tourists had been accommodated. As busload after busload of people roared up before the observatory, we saw that this was an indirect dismissal. It would be dawn before our turn came, if, indeed, they chose to accept us at all.

The guard eyed us and our clothes with sullen distaste. Though it was freezing cold upon the mountain, it was plain we were not welcome in the inn that catered to the tourists. Reluctantly, with a few coins from our little store of change we purchased a bit of chocolate. We looked at each other. Wearily and without a word, we turned and began our long descent through the dark. It would take many hours; nor were we sustained by having seen the shining planet upon which our hopes were fixed.

This was my first experience of the commercial side of outer space, and though I now serve upon a committee that encourages the young in a direction once denied me, I feel that this youthful experience contributed to a certain growing introspection and curiosity about the relationship of science to the world about it.

Something was seriously wrong upon that mountain and among the wise men who flourished there. Knowledge, I had learned in the bleak wind by the shut door, was not free, and many to whom that observatory was only a passing curiosity had easier access to it than we who had climbed painfully for many hours. My memory is from the far days of the twenties, and I realize that we now beckon enticingly to the youth interested in space where before we ignored him. I still have an uncomfortable feeling, however, that it is the circumstances, and not the actors, which have changed. I remain oppressed by the thought that the venture into space is meaningless unless it coincides with a certain interior expansion, an ever growing universe within, to correspond with the far flight of the galaxies our telescopes follow from without.

Upon that desolate peak my mind had turned finally inward. It is from that domain, that inner sky, that I choose to speak—a world of dreams, of light and darkness, that we will never escape, even on the far edge of Arcturus. The inward skies of man will accompany him across any void upon which he ventures and will be with him to the end of time. There is just one way in which that inward world differs from outer space. It can be more volatile and mobile, more terrible and impoverished, yet withal more ennobling in its self-consciousness, than the universe that gave it birth. To the educators of this revolutionary generation, the transformations we may induce in that inner sky loom in at least equal importance with the work of those whose goals are set beyond the orbit of the moon.

No one needs to be told that different and private worlds exist in the heads of men. But in a day when some men are listening by radio telescope to the rustling of events at the ends of the universe, the universe of others consists of hopeless poverty amidst the filthy garbage of a city lot. A taxi man I know thinks the stars are just "up there," and that as soon as our vehicles are perfected we can all take off like crowds of summer tourists to Cape Cod. This man expects, and I fear has been encouraged to expect, that such flights will solve the population problem. Again, while I was sitting one night with a poet friend watching a great opera performed in a tent under arc lights, the poet took my arm and pointed silently. Far up, blundering out of the night, a huge Cecropia moth swept past from light to light over the posturings of the actors.

"He doesn't know," my friend whispered excitedly. "He's passing through an alien universe brightly lit but invisible to him. He's in another play; he doesn't see us. He doesn't know. Maybe it's happening right now to us. Where are we? Whose is the *real* play?"

Between the universe of the moth and the poet, I sat confounded. My mind went back to the heads of alabaster that the kings of the old Egyptian Empire sought to endow with eternal life, replacing thus against accident their own frail and perishable brains for the passage through eternity. The Pharaohs, like the moth among the arc lights, had been entranced by the flaming journey of the sun. Some had even constructed, hopefully, their own solar boats. Perhaps, I thought, those boats symbolized the frail vessel of which Plato was later to speak—that vessel on which to risk the voyage of life, or, rather, eternity, which was inevitably man's compulsive interest. As for me, I had come to seek wisdom no longer upon the improvised rafts of proud philosophies. I had seen the moth burn in its passage through the light. I had seen all the vessels fail but one—that word which Plato sought, and which none could long identify or hold.

There *was* a real play, but it was a play in which man was destined always to be a searcher, and it would be his true nature he would seek. The fragile vessel was himself, and not among the stars upon the mountain. Was not that what Plotinus had implied? Then if a man were to write further, I considered, he would write of that—of the last things.

2

Several years ago, a man in a small California town suffered an odd accident. The accident itself was commonplace. But the psychological episode accompanying it seems so strange that I recount it here. I had been long engaged upon a book I was eager to finish. As I walked, abstracted and alone, toward my office one late afternoon, I caught the toe of my shoe in an ill-placed drain. Some trick of mechanics brought me down over the curb with extraordinary violence. A tremendous crack echoed in my ears. When I next opened my eyes I was lying face down on the sidewalk. My nose was smashed over on one side. Blood from a gash on my forehead was cascading over my face.

Reluctantly I explored further, running my tongue cautiously about my mouth and over my teeth. Under my face a steady rivulet of blood was enlarging to a bright red pool on the sidewalk. It was then, as I peered nearsightedly at my ebbing substance there in the brilliant sunshine, that a surprising thing happened. Confusedly, painfully, indifferent to running feet and the anxious cries of witnesses about me, I lifted a wet hand out of this welter and murmured in compassionate concern, "Oh, don't go. I'm sorry, I've done for you."

The words were not addressed to the crowd gathering about me. They were inside and spoken to no one but a part of myself. I was quite sane, only it was an oddly detached sanity, for I was addressing blood cells, phagocytes, platelets, all the crawling, living, independent wonder that had been part of me and now, through my folly and lack of care, were dying like beached fish on the hot pavement. A great wave of passionate contrition, even of adoration, swept through my mind, a sensation of love on a cosmic scale, for mark that this experience was, in its way, as vast a catastrophe as would be that of a galaxy consciously suffering through the loss of its solar systems.

I was made up of millions of these tiny creatures, their toil, their sacrifices, as they hurried to seal and repair the rent fabric of this vast being whom they had unknowingly, but in love, compounded. And I, for the first time in my mortal existence, did not see these creatures as odd objects under the microscope. Instead, an echo of the force that moved them came up from the deep well of my being and flooded through the shaken circuits of my brain. I was their galaxy, their creation. For the first time, I loved them consciously, even as I was plucked up and carried away by willing hands. It seemed to me then, and does now in retrospect, that I had caused to the universe I inhabited as many deaths as the explosion of a supernova in the cosmos.

Weeks later, recovering, I paid a visit to the place of the accident. A faint discoloration still marked the sidewalk. I hovered over the spot, obscurely troubled. They were gone, utterly destroyed—those tiny beings—but the entity of which they had made a portion still persisted. I shook my head, conscious of the brooding mystery that the poet Dante impelled into his great line: "the love that moves the sun and other stars."

The phrase does not come handily to our lips today. For a century we have chosen to talk continuously about the struggle for existence, about man, the brawling half-ape and bestial fighter. We have explored with wavering candles the dark cellars of our subconscious and been appalled by the faces we have encountered there. It will do no harm, therefore, if we choose to examine the history of that great impulse—love, compassion, call it what one will—which, however discounted in our time, moved the dying Christ on Golgotha with a power that has reached across two thousand weary years.

"The conviction of wisdom," wrote Montaigne in the sixteenth century, "is the plague of man." Century after century, humanity studies itself in the mirror of fashion, and ever the mirror gives back distortions, which for the moment impose themselves upon man's real image. In one period we believe ourselves governed by immutable laws; in the next, by chance. In one period angels hover over our birth; in the following time we are planetary waifs, the product of a meaningless and ever altering chemistry. We exchange halos in one era for fangs in another. Our religious and philosophical conceptions change so rapidly that the theological and moral exhortations of one decade become the wastepaper of the next epoch. The ideas for which millions yielded up their lives produce only bored yawns in a later generation.

"We are, I know not how," Montaigne continued,

"double in ourselves, so that what we believe we disbelieve, and cannot rid ourselves of what we condemn."

This complex, many-faceted, self-conscious creature now examines himself in the mirror held up to him by the modern students of prehistory. Increasingly he asks of the bony fragments recovered from pre-ice-age strata, not whether they are related to himself, but what manner of creature they proclaim us to be. Of the answer that may come up from underground we are all too evidently afraid. There are even those who have dared prematurely to announce the verdict. "Look," they say, "at the dark instincts that drive you. Look deep into your bloody, fossil, encrusted hearts. Then you will know man. You will know him from the caves to the Berlin wall. Thus he is and thus he will remain. It is written in his bones."

Yet the moment the words are said and documented, either the data are seized upon to give ourselves a fearsome picture to delight and excuse the black side of our natures or, strangely, even beautifully, the picture begins to waver and to change. St. Francis of the birds broods by the waters; Gilbert White of Selborne putters harmlessly with the old pet tortoise in his garden. Ishi, the primitive gentle philosopher, steps real as life from the Sierra forest—the idyllic man denounced as an invention of Rousseau's, yet the product of a world more primitive than black Africa today.

"Double in ourselves" we are, said Montaigne. Now with that doubleness in mind let us look once more into the fossil past, full into the hollow sockets of the halfmen from whom we sprang. Their bones are known; their remains have been turning up for over a century in almost every area of the Old World land mass. They have been found in the caves and gravels of ice age Europe, in the cemented breccias of deposits near Peking, in Asian coastal isles like Java, shaken at intervals by turbulent volcanoes. They have been found, as well, in the high uplands of eastern Africa and in the grottoes of the Holy Land.

Nevertheless, the faces of our ancestors remain forever unknown to us even as they stare from the illustrations of the poorest and most obscure textbook. The color of their skins is lost, the texture of their hair unknown, the expression of their once living features is as masked as those of the anonymous cadaver that represents collective humanity in the pages of medical textbooks. It is the same gray anonymity in which man's formidable enemy, the saber-toothed tiger, is lost, or even the dinosaur.

In the case of man, the representations are particularly ungratifying. Man is a creature volatile of expression, and across his features in a day may flow happiness and remorse, rage and charity. Individually, as on a modern street, one should be able to sight the sly, the brutal, and the benignant. If, in the world of fossils, however, we seek the soul of man himself, we are forced to draw it from the empty sockets of skulls or the representations of artists quick to project their own conceptions of the past upon the indifferent dead.

It is man's folly, as it is perhaps a sign of his spiritual aspirations, that he is forever scrutinizing and redefining himself. A mole, so far as we can determine, is content with its dim world below the grass roots, a snow leopard with being what he is—a drifting ghost in a blizzard. Man, by contrast, is marked by a restless inner eye, which, in periods of social violence, such as characterize our age, grows clouded with anxiety. There are times when our bodies seem to waver from within and bulge lumpishly with the shape of contending forces.

There is danger as well as wisdom, however, in such self-scrutiny. Man, unlike the lower creatures locked safely within their particular endowed natures, possesses freedom. He can define and redefine his own humanity, his own conception of himself. In so doing, he may give wings to the spirit or reshape himself into something more genuinely bestial than any beast of prey obeying its own nature. In this ability to take on the shape of his own dreams, man extends beyond visible nature into another and stranger realm. It is part of each person's individual evolutionary status that he possesses this power in unequal degrees.

Few of us can be saints; few of us are total monsters. To the degree that we let others project upon us erroneous or unbalanced conceptions of our natures, we may unconsciously reshape our own image to less pleasing forms. It is one thing to be "realistic," as many are fond of saying, about human nature. It is another thing entirely to let that consideration set limits to our spiritual aspirations or to precipitate us into cynicism and despair. We are protean in many things, and stand between extremes. There is still great room for the observation of John Donne, made over three centuries ago, however, that "no man doth refine and exalt Nature to the heighth it would beare."

As one surveys the artistic conceptions of the past, whether sculptured or drawn, one frequently encounters an adenoidal, open-mouthed brute with a club representing Neanderthal man. Then, by contrast, we encounter a neatly groomed model of Peking man, looking as clear-eyed and intelligent as a broker on his way to the Stock Exchange. Something is obviously wrong here. The well-groomed Peking specimen belongs on the same anatomical level as Pithecanthropus, sometimes represented in older illustrations as possessing snarling fangs. The fangs are a figment of the artist's imagination. They have been stolen from our living relative, the gorilla. The mispictured adenoidal moron with the club is known to have buried his dead with offerings, and to have cared for the injured and maimed among his kind.

Men are subjects of society. It is true that they carry bits and pieces of their past about with them, but they also covertly examine in the social mirror of their minds the way they look. Thus there is a quality of illusion about all of us. Emerson knew this well when he asked, in one of his more profound moments, "Why do men feel that the natural history of man has never been written, but he is always leaving behind what you have said of him, and it becomes old and books of metaphysics worthless?"

This comment of Emerson's is perhaps one of the most difficult pieces of wisdom that man has to learn. We are inclined to visualize our psychological make-up as fixed—as something bestowed upon the first man. In pre-evolutionary times, the human mind, with its reason, its conscience, its free will, was regarded as divinely and immediately created in the human organism just as it stands today.

With the rise of Darwinian evolution in the mid-nineteenth century, the concept of the stably endowed species correctly gave way to the notion of man and other animal forms as transient, imperfect, forever moving from one set of conditions to another. "Cosmic nature," wrote Thomas Huxley, Darwin's colleague and defender, "is no school of virtue. . . . For his successful progress as far as the savage state, man has been largely indebted to those qualities which he shares with the ape and the tiger."

No intelligent person today, surveying the low skull vault and heavy brow ridges of fossil man, can deny that man has changed through the aeons of prehistory, however difficult may seem the road he has traveled. Natural selection has undoubtedly played a leading role in that process. Here we must proceed with care, if we are not to fall into fallacious reasoning. Otherwise we will emerge from our survey of the past with another set of stereotypes as to the nature of man, which may well prove to be just as rigid and dogmatic as those developed in pre-evolutionary thought—stereotypes that have been thrust forward even today as evidence of man's bestial nature.

Man's altruistic and innately co-operative character has brought him along the road to civilization far more than the qualities of the ape and tiger of Huxley's analysis. These are bad metaphors at best. The ape is a largely inoffensive social animal, the tiger a solitary, carnivorous hunter. To lump them in a comparison with man is spectacular but confusing. As for the fearful war of nature painted by the early evolutionists and symbolized by the tiger, we know today that even the great carnivores exist, normally, in balance with their prey. When satiated and not involved in the hunt, they may stroll scarcely noticed among the herd creatures they stalk.

Some members of the Darwinian circle could only conceive of man achieving his high intellect through the heavy selection of incessant war. Today we know that early man was small and scant in numbers and that most of his efforts must have been given over to food-getting rather than conflict. This is not to minimize his destructive qualities, but his long-drawn-out, helpless childhood, during which his growing brain matured, could only have flourished in the safety of a stable family organization—groups marked by altruistic and long-continued care of the young.

The nineteenth-century evolutionists, and many philosophers still today, are obsessed by struggle. They try to define natural selection in one sense only—something that Darwin himself avoided. They ignore all man's finer qualities—generosity, self-sacrifice, universe-searching wisdom—in the attempt to enclose him in the small capsule that contained the brain of protoman. Such writers often fail to explore man's growing sense of beauty, the language that has opened and defined his world, the little gifts he came to lay beside his dead.

None of these acts could have been prophesied before man came. They reveal something other than what the pure materialist would be able to draw out of the dark concourse of matter before the genuine emergence of these novel human phenomena into time. There is no definition or description of man possible by reducing him to ape or tree shrew. Once, it is true, the shrew contained him, but he is gone. He has broken from the opened seed pod of the prehominid brain, a thistledown now drifting toward the empty spaces of the universe. He is full of the lights and visions—yes, and the fearful darknesses—of the next age of man.

The world we now know is open-ended, unpredictable. Man has partially domesticated himself; in this lies the story of his strange nature, of that love which transcends the small Darwinian matters of tribal cooperation and safety. For man, be it noted, can love the music of Ariel's isle, or, in his heart, that ideal city of the Greeks which is not and yet is forever.

The law of selection that acts upon living creatures in the wild is frequently repressive. A coat color a little off tone and visible, a variation in instinct, may make for death. The powerful creative surge from the underdarkness of nature is held in check, awaiting, perhaps, a season that never comes; the white stag is struck down by the hunter. It is this unending struggle that those who would picture man from the beginning as a monster of terror would delineate—the man with the stone striking down in barbaric rage, not only his game, but his brother and his son.

Natural selection is real but at the same time it is a shifting chimera, less a "law" than making its own law from age to age. Let us see, before we approach what I shall call domesticated man, what mutual aid can mean in the life of a European sea bird, the common tern. This bird lacks the careful concealing coloration of some of nature's species. It is variable in matters of egg form and nest shape. Capricious deviation in all these features prevails among the terns. The conformist pressures of natural selection have here given way to the creative forces of random mutation. The potential hidden in nature has flowered into a greater variety of behavior. Thus, what we call natural selection, "the war of nature," can either enclose living creatures in specialized prisons or, on occasion, open amazing doorways into unsuspected worlds. Even such a lowly relative of man as the existing lemur Propithecus, which lives in groups, may exhibit marked individual variation, because these animals recognize and behave differently toward each other. Conformity has here given way to selective pressure for at least limited physical diversity and corresponding individuality of behavior.

Though the case of man is complicated, it seems evident that just such a remarkable doorway opened when man, as a social animal, fell under selective forces that no longer severely channeled the nature of his mind or the minds of his aberrant offspring. Through language, this creature could communicate his dreams around the cave fires. Inevitably, a great wealth of intellectual diversity, and consequent selective mating, based upon mutual attraction, would emerge from the dark storehouse of nature. The cruel and the gentle would sit at the same fireside, dreaming already in the Stone Age the different dreams they dream today.

The visionary was already awaiting the eternal city; the gifted musician sat hearing in his brain sounds that did not yet exist. All waited upon and yet possessed, in some dim way, the future in their heads. Abysmal darkness and great light lay invisibly about their camps. The phantom cities of the far future awaited latent talents for which, in that unspecialized time, there was no name.

Above all, some of them, a mere handful in any generation perhaps, loved—they loved the animals about them, the song of the wind, the soft voices of women. On the flat surfaces of cave walls the three dimensions of the outside world took animal shape and form. Here—not with the ax, not with the bow—man fumbled at the door of his true kingdom. Here, hidden in times of trouble behind silent brows, against the man with the flint, waited St. Francis of the birds—the lovers, the men who are still forced to walk warily among their kind.

3

I am middle-aged now, and like the Egyptian heads of buried stone, or like the gentle ones who came before me, I am resigned to wait out man's lingering barbarity. I have walked much to the sea, not knowing what I seek. The west headland I visit is always boiling, even on calm days. Spume leaps up from the sea caverns of buried reefs and the blue and purple of the turbulent waters are roiled and twisted with clashing and opposed currents. I go there frequently and sit for hours on an old whiskey crate half-buried in the sand.

Staring into those uncertain and treacherous waters with their unexpected and lifting apparitions is like looking into the future. You can see its forces constantly gathering, expending themselves, streaming away and streaming back, contorting or violently lifting into huge and grotesque shapes. The meaning escapes one, but day after day the harpy gulls scream and mew over it and the crabs scuttle like spiders along its edge, waving threatening pincers.

But I wander.

On one occasion, there was just this broken crate in the sand, myself, and the sea—and then this other. I only became aware of him after several days had passed. I first encountered him when I had ventured at low tide up to the verge of the reef beyond which burst that leaping, spouting thunder, which, in my isolated wanderings, I had come to conceive of as containing the future. As I reached the flat, slippery stones over which passed a constant surf, I saw a gray wing tilt upward and move a few feet farther on. It was a big gray-backed gull, who slid quietly down again amidst the encrusted sea growth. He moved just enough, out of old and wise judgment, to keep me at arm's length, no more. He was no longer with his kind, hovering and mewing over the outer rock masses of a dubious future. He had a space of his own on the last edge of the present. He fed there upon such things as the sea brought. He was old and he rested, if one could be said to rest amidst such waters.

I disturbed him once by coming closer, whereupon he rose and tilted slightly in the blast from over the reef. If I did not move, neither did he. Since I am not one to go rushing over dangerous crevices, we achieved, after some days, a dignified relationship. We were both gray, and disinclined toward a future that had come to have little meaning to either of us. We stood or sat a little apart and ignored each other, being, after all, creatures diverse.

Every morning when I came he was there. He was growing thinner, but he still rose at my coming and hovered low upon his great seagoing wings. Then I would seek my box and he would swoop back to the little space that contained his last of life. I came to look for this bird as though we shared some sane, enormously simple secret amidst a little shingle of hard stones and broken beach.

After several days he was gone. A sector of my own

life had been sheared away with his going. I shied a stone uncertainly toward the still-spouting future. Nothing came of it; no hand reached out, no shape emerged. The only rational shape had been that aged gull, too wise to venture more than a tilting wing's length upward in such air. Finally, the extremest edge of his space had hesitantly touched mine. Neither of us had much farther to go, and the harsh simplicity of it was somehow appropriate and gratifying. A little salt-washed rock had contained us both.

Here, I thought, is where I shall abide my ending, in the mind at least. Here where the sea grinds coral and bone alike to pebbles, and the crabs come in the night for the recent dead. Here where everything is transmuted and transmutes, but all is living or about to live.

It was here that I came to know the final phase of love in the mind of man—the phase beyond the evolutionists' meager concentration upon survival. Here I no longer cared about survival—I merely loved. And the love was meaningless, as the harsh Victorian Darwinists would have understood it or even, equally, those harsh modern materialists of whom Lord Dunsany once said: "It is very seldom that the same man knows much of science, and about the things that were known before ever science came."

I felt, sitting in that desolate spot upon my whiskey crate, a love without issue, tenuous, almost disembodied. It was a love for an old gull, for wild dogs playing in the surf, for a hermit crab in an abandoned shell.

It was a love that had been growing through the unthinking demands of childhood, through the pains and rapture of adult desire. Now it was breaking free, at last, of my worn body, still containing but passing beyond those other loves. Now, at last, it was truly "the bright stranger, the foreign self," of which Emerson had once written.

Through shattered and receding skulls, growing ever smaller behind us in the crannies of a broken earth, a stranger had crept and made his way. But precisely how he came, and what might be his destiny, except that it is not wholly of our time or this our star, we do not know.

Perhaps it is always the destined role of the compassionate to be strangers among men. To fail and pass, to fail and come again. For the seed of man is thistledown, and a puff of breath may govern it, or a word from a poet torment it into greatness. There are few among us who can notice the passage of a moth's wing across an opera tent at midnight and ask ourselves, "Whose is the real play?"

I had turned to the young man who spoke those words as to one whose eye reached farther than the giant lens upon the mountain in my youth. Before us had seemed to stretch the infinite pathways of space down which, like the questing moth, it was henceforth man's doom to wander. But the void had become to me equally an interior void—the void of our own minds—a sea as infinite as the one before which I had been meditating.

Amidst the fall of waters on that desolate shore I

watched briefly an exquisitely shaped jellyfish pumping its little umbrella sturdily along only to subside with the next wave on the strand. "Love makyth the lover and the living matters not," an old phrase came hesitantly to my lips. We would win, I thought steadily, if not in human guise then in another, for love was something that life in its infinite prodigality could afford. It was the failures who had always won, but by the time they won they had come to be called successes. This is the final paradox, which men call evolution.

NINE

The Innocent Fox

Only to a magician is the world forever fluid, infinitely mutable and eternally new. Only he knows the secret of change, only he knows truly that all things are crouched in eagerness to become something else, and it is from this universal tension that he draws his power. —PETER BEAGLE

SINCE man first saw an impossible visage staring upward from a still pool, he has been haunted by meanings—meanings felt even in the wood, where the trees leaned over him, manifesting a vast and living presence. The image in the pool vanished at the touch of his finger, but he went home and created a legend. The great trees never spoke, but man knew that dryads slipped among their boles. Since the red morning of time it has been so, and the compulsive reading of such manuscripts will continue to occupy man's attention long after the books that contain his inmost thoughts have been sealed away by the indefatigable spider.

Some men are daylight readers, who peruse the ambiguous wording of clouds or the individual letter shapes of wandering birds. Some, like myself, are librarians of the night, whose ephemeral documents consist of root-inscribed bones or whatever rustles in thickets upon solitary walks. Man, for all his daylight activities, is, at best, an evening creature. Our very addiction to the day and our compulsion, manifest through the ages, to invent and use illuminating devices, to contest with midnight, to cast off sleep as we would death, suggest that we know more of the shadows than we are willing to recognize. We have come from the dark wood of the past, and our bodies carry the scars and unhealed wounds of that transition. Our minds are haunted by night terrors that arise from the subterranean domain of racial and private memories.

Lastly, we inhabit a spiritual twilight on this planet. It is perhaps the most poignant of all the deprivations to which man has been exposed by nature. I have said *deprivation,* but perhaps I should, rather, maintain that this feeling of loss is an unrealized anticipation. We imagine we are day creatures, but we grope in a lawless and smoky realm toward an exit that eludes us. We appear to know instinctively that such an exit exists.

I am not the first man to have lost his way only to find, if not a gate, a mysterious hole in a hedge that a child would know at once led to some other dimension at the world's end. Such passageways exist, or man would not be here. Not for nothing did Santayana once contend that life is a movement from the forgotten into the unexpected.

As adults, we are preoccupied with living. As a consequence, we see little. At the approach of age some men look about them at last and discover the hole in the hedge leading to the unforeseen. By then, there is frequently no child companion to lead them safely through. After one or two experiences of getting impaled on thorns, the most persistent individual is apt to withdraw and to assert angrily that no such opening exists.

My experience has been quite the opposite, but I have been fortunate. After several unsuccessful but tantalizing trials, which I intend to disclose, I had the help, not of a child, but of a creature—a creature who, appropriately, came out of a quite unremarkable and prosaic den. There was nothing, in retrospect, at all mysterious or unreal about him. Nevertheless, the creature was baffling, just as, I suppose, to animals, man himself is baffling.

2

An autumn midnight in 1967 caught me staring idly from my study window at the attic cupola of an old Victorian house that loomed far above a neighboring grove of trees. I suppose the episode happened just as I had grown dimly aware, amidst my encasing cocoon of books and papers, that something was missing from my life. This feeling had brought me from my desk to peer hopelessly upon the relentless advance of suburban housing. For years, I had not seen anything from that particular window that did not spell the death of something I loved.

Finally, in blundering, good-natured confidence, the last land tortoise had fallen a victim to the new expressway. None of his kind any longer came to replace him. A chipmunk that had held out valiantly in a drainpipe on the lawn had been forced to flee from the usurping rats that had come with the new supermarket. A parking lot now occupied most of the view from the window. I was a man trapped in the despair once alluded to as the utterly hopeless fear confined to moderns—that no miracles can ever happen. I considered, as I tried to will myself away into the attic room far above the trees, the wisdom of a search, a search unlikely to yield tangible results.

Since boyhood I have been charmed by the unexpected and the beautiful. This was what had led me originally into science, but now I felt instinctively that something more was needed—though what I needed verged on a miracle. As a scientist, I did not believe in miracles, though I willingly granted the word broad latitudes of definition.

My whole life had been unconsciously a search, and the search had not been restricted to the bones and stones of my visible profession. Moreover, my age could allow me folly; indeed, it demanded a boldness that the young frequently cannot afford. All I needed to do was to set forth either mentally or physically, but to where escaped me.

At that instant the high dormer window beyond the

trees blazed as blue as a lightning flash. As I have remarked, it was midnight. There was no possibility of reflection from a street lamp. A giant bolt of artificial lightning was playing from a condenser, leaping at intervals across the interior of the black pane in the distance. It was the artificial lightning that only one or several engineers with unusual equipment could produce.

Now the old house was plebeian enough. Rooms were rented. People of modest middle-class means lived there, as I was to learn later. But still, in the midmost of the night, somebody or some group was engaged in that attic room upon a fantastic experiment. For, you see, I spied. I spied for nights in succession. I was bored, I was sleepless, and it pleased me to think that the mad scientists, as I came to call them, were engaged, in their hidden room, upon some remarkable and unheard-of adventure.

Why else would they be active at midnight, why else would they be engaged for a brief hour and then extinguish the spark? In the next few days I trained high-powered field glasses upon the window, but the blue bolt defeated me, as did the wavering of autumn boughs across the distant roof. I could only believe that science still possessed some of its old, mad fascination for a mind outside the professional circle of the great laboratories. Perhaps, I thought eagerly, there was a fresh intelligence groping after some secret beyond pure technology. I thought of the dreams of Emerson and others when evolution was first anticipated but its mechanisms remained a mystery entangled with the first galvanic batteries. Night after night, while the leaves thinned and the bolt leaped at its appointed hour, I dreamed, staring from my window, of that coruscating arc revivifying flesh or leaping sentient beyond it into some unguessed state of being. Only for such purposes, I thought, would a man toil in an attic room at midnight.

I began unconsciously to hang more and more upon that work of which, in reality, I knew nothing. It sustained me in my waking hours when the old house, amidst its yellowing leaves, assumed a sleepy and inconsequential air. For me, it had restored wonder and lifted my dreams to the height they had once had when, as a young student, I had peeped through the glass door of a famous experimenter's laboratory. I no longer read. I sat in the darkened study and watched and waited for the unforeseen. It came in a way I had not expected.

One night the window remained dark. My powerful glasses revealed only birds flying across the face of the moon. A bat fluttered about the tessellated chimney. A few remaining leaves fell into the dark below the roofs.

I waited expectantly for the experiment to be resumed. It was not. The next night it rained violently. The window did not glow. Leaves yellowed the wet walks below the street lamps. It was the same the next night and the next. The episode, I came to feel, peering ruefully from my window, was altogether too much like science itself—science with its lightning bolts, its bubbling retorts, its elusive promises of perfection. All too frequently the dream ended in a downpour of rain and leaves upon wet walks. The men involved had a way, like my mysterious neighbors, of vanishing silently and leaving, if anything at all, corroding bits of metal out of which no one could make sense.

I had once stood in a graveyard that was a great fallen city. It was not hard to imagine another. After watching fruitlessly at intervals until winter was imminent, I promised myself a journey. After all, there was nothing to explain my disappointment. I had not known for what I was searching.

Or perhaps I did know, secretly, and would not admit it to myself: I wanted a miracle. Miracles, by definition, are without continuity, and perhaps my rooftop scientist had nudged me in that direction by the uncertainty of his departure. The only thing that characterizes a miracle, to my mind, is its sudden appearance and disappearance within the natural order, although, strangely, this loose definition would include each individual person. Miracles, in fact, momentarily dissolve the natural order or place themselves in opposition to it. My first experience had been only a tantalizing expectation, a hint that I must look elsewhere than in retorts or coiled wire, however formidable the powers that could be coerced to inhabit them. There was magic, but it was an autumnal, sad magic. I had a growing feeling that miracles were particularly concerned with life, with the animal aspect of things.

Just at this time, and with my thoughts in a receptive mood, a summons came that made it necessary for me to make a long night drive over poor roads through a dense forest. As a subjective experience, which it turned out to be, I would call it a near approach to what I was seeking. There was no doubt I was working further toward the heart of the problem. The common man thinks a miracle can just be "seen" to be reported. Quite the contrary. One has to be, I was discovering, reasonably sophisticated even to *perceive* the miraculous. It takes experience; otherwise, more miracles would be encountered.

One has, in short, to refine one's perceptions. Lightning bolts observed in attics, I now knew, were simply raw material, a lurking extravagant potential in the cosmos. In themselves, they were merely powers summoned up and released by the human mind. Wishing would never make them anything else and might make them worse. Nuclear fission was a ready example. No, a miracle was definitely something else, but that I would have to discover in my own good time.

Preoccupied with such thoughts, I started my journey of descent through the mountains. For a long time I was alone. I followed a road of unexpectedly twisting curves and abrupt descents. I bumped over ruts, where I occasionally caught the earthly starshine of eyes under leaves. Or I plunged at intervals into an impenetrable gloom buttressed by the trunks of huge pines.

After hours of arduous concentration and the sudden crimping of the wheel, my eyes were playing tricks with me. It was time to stop, but I could not afford to stop. I shook my head to clear it and blundered on. For a long time, in this confined glen among the mountains,

I had been dimly aware that something beyond the reach of my headlights, but at times momentarily caught in their flicker, was accompanying me.

Whatever the creature might be, it was amazingly fleet. I never really saw its true outline. It seemed, at times, to my weary and much-rubbed eyes, to be running upright like a man, or, again, its color appeared to shift in a multiform illusion. Sometimes it seemed to be bounding forward. Sometimes it seemed to present a face to me and dance backward. From weary consciousness of an animal I grew slowly aware that the being caught momentarily in my flickering headlights was as much a shapeshifter as the wolf in a folk tale. It was not an animal; it was a gliding, leaping mythology. I felt the skin crawl on the back of my neck, for this was still the forest of the windigo and the floating heads commemorated so vividly in the masks of the Iroquois. I was lost, but I understood the forest. The blood that ran in me was not urban. I almost said not human. It had come from other times and a far place.

I slowed the car and silently fought to contain the horror that even animals feel before the disruption of the natural order. But was there a natural order? As I coaxed my lights to a fuller blaze I suddenly realized the absurdity of the question. Why should life tremble before the unexpected if it had not already anticipated the answer? There was no order. Or, better, what order there might be was far wilder and more formidable than that conjured up by human effort.

It did not help in the least to make out finally that

the creature who had assigned himself to me was an absurdly spotted dog of dubious affinities—nor did it help that his coat had the curious properties generally attributable to a magician. For how, after all, could I assert with surety what shape this dog had originally possessed a half mile down the road? There was no way of securing his word for it.

The dog was, in actuality, an illusory succession of forms finally, but momentarily, frozen into the shape "dog" by me. A word, no more. But as it turned away into the night how was I to know it would remain "dog"? By experience? No, it had been picked by me out of a running weave of colors and faces into which it would lapse once more as it bounded silently into the inhuman, unpopulated wood. We deceive ourselves if we think our self-drawn categories exist there. The dog would simply become once more an endless running series of forms, which would not, the instant I might vanish, any longer know themselves as "dog."

By a mental effort peculiar to man, I had wrenched a leaping phantom into the flesh "dog," but the shape could not be held, neither his nor my own. We were contradictions and unreal. A nerve net and the lens of an eye had created us. Like the dog, I was destined to leap away at last into the unknown wood. My flesh, my own seemingly unique individuality, was already slipping like flying mist, like the colors of the dog, away from the little parcel of my bones. If there was order in us, it was the order of change. I started the car again, but I drove on chastened and unsure. Somewhere something was running and changing in the haunted wood. I knew no more than that. In a similar way, my mind was leaping and also changing as it sped. That was how the true miracle, my own miracle, came to me in its own time and fashion.

3

The episode occurred upon an unengaging and unfrequented shore. It began in the late afternoon of a day devoted at the start to ordinary scientific purposes. There was the broken prow of a beached boat subsiding in heavy sand, left by the whim of ancient currents a long way distant from the shifting coast. Somewhere on the horizon wavered the tenuous outlines of a misplaced building, growing increasingly insubstantial in the autumn light.

After my companions had taken their photographs and departed, their persistent voices were immediately seized upon and absorbed by the extending immensity of an incoming fog. The fog trailed in wisps over the upthrust ribs of the boat. For a time I could see it fingering the tracks of some small animal, as though engaged in a belated dialogue with the creature's mind. The tracks crisscrossed a dune, and there the fog hesitated, as though puzzled. Finally, it approached and enwrapped me, as though to peer into my face. I was not frightened, but I also realized with a slight shock that I was not intended immediately to leave.

I sat down then and rested with my back against the

overturned boat. All around me the stillness intensified and the wandering tendrils of the fog continued their search. Nothing escaped them.

The broken cup of a wild bird's egg was touched tentatively, as if with meaning, for the first time. I saw a sand-colored ghost crab, hitherto hidden and immobile, begin to sidle amidst the beach grass as though imbued suddenly with a will derived ultimately from the fog. A gull passed high overhead, but its cry took on the plaint of something other than itself.

I began dimly to remember a primitive dialogue as to whether God is a mist or merely a mist maker. Since a great deal of my thought has been spent amidst such early human and, to my mind, not outworn speculations, the idea did not seem particularly irrational or blasphemous. How else would so great a being, assuming his existence, be able thoroughly to investigate his world, or, perhaps, merely a world that he had come upon, than as he was now proceeding to do?

I closed my eyes and let the tiny diffused droplets of the fog gently palpate my face. At the same time, by some unexplained affinity, I felt my mind drawn inland, to pour, smoking and gigantic as the fog itself, through the gorges of a neighboring mountain range.

In a little shaft of falling light my consciousness swirled dimly over the tombstones of a fallen cemetery. Something within me touched half-obliterated names and dates before sliding imperceptibly onward toward an errand in the city. That errand, whatever its purpose, perhaps because I was mercifully guided away from the future, was denied me.

As suddenly as I had been dispersed I found myself back among the boat timbers and the broken shell of something that had not achieved existence. "I am the thing that lives in the midst of the bones"—a line from the dead poet Charles Williams persisted obstinately in my head. It was true. I was merely condensed from that greater fog to a smaller congelation of droplets. Vague and smoky wisplets of thought were my extensions.

From a rack of bone no more substantial than the broken boat ribs on the beach, I was moving like that larger, all-investigating fog through the doorways of the past. Somewhere far away in an inland city the fog was transformed into a blizzard. Nineteen twenty-nine was a meaningless date that whipped by upon a flying newspaper. The blizzard was beating upon a great gate marked St. Elizabeth's. I was no longer the blizzard. I was hurrying, a small dark shadow, up a stairway beyond which came a labored and importunate breathing.

The man lay back among the pillows, wracked, yellow, and cadaverous. Though I was his son he knew me only as one lamp is briefly lit from another in the windy night. He was beyond speech, but a question was there, occupying the dying mind, excluding the living, something before which all remaining thought had to be mustered. At the time I was too young to understand. Only now could the hurrying shadow drawn from the wrecked boat interpret and relive the question. The starving figure on the bed was held back from death only by a magnificent heart that would not die.

I, the insubstantial substance of memory, the dispersed droplets of the ranging fog, saw the man lift his hands for the last time. Strangely, in all that ravished body, they alone had remained unchanged. They were strong hands, the hands of a craftsman who had played many roles in his life: actor, laborer, professional runner. They were the hands of a man, indirectly of all men, for such had been the nature of his life. Now, in a last lucid moment, he had lifted them up and, curiously, as though they belonged to another being, he had turned and flexed them, gazed upon them unbelievingly, and dropped them once more.

He, too, the shadow, the mist in the gaping bones, had seen these seemingly untouched deathless instruments rally as though with one last purpose before the demanding will. And I, also a shadow, come back across forty years, could hear the question at last. "Why are you, my hands, so separate from me at death, yet still to be commanded? Why have you served me, you who are alive and ingeniously clever?" For here he turned and contemplated them with his old superb steadiness. "What has been our partnership, for I, the shadow, am going, yet you of all of me are alive and persist?"

I could have sworn that his last thought was not of himself but of the fate of the instruments. He was outside, he was trying to look into the secret purposes of things, and the hands, the masterful hands, were the only purpose remaining, while he, increasingly without center, was vanishing. It was the hands that contained his last conscious act. They had been formidable in life.

In death they had become strangers who had denied their master's last question.

Suddenly I was back under the overhang of the foundered boat. I had sat there stiff with cold for many hours. I was no longer the extension of a blizzard beating against immovable gates. The year of the locusts was done. It was, instead, the year of the mist maker that some obscure Macusi witch doctor had chosen to call god. But the mist maker had gone over the long-abandoned beach, touching for his inscrutable purposes only the broken shell of the nonexistent, only the tracks of a wayward fox, only a man who, serving the mist maker, could be made to stream wispily through the interstices of time.

I was a biologist, but I chose not to examine my hands. The fog and the night were lifting. I had been far away for hours. Crouched in my heavy sheepskin I waited without thought as the witch doctor might have waited for the morning dispersion of his god. Finally, the dawn began to touch the sea, and then the worn timbers of the hulk beside which I sheltered reddened just a little. It was then I began to glimpse the world from a different perspective.

I had watched for nights the great bolts leaping across the pane of an attic window, the bolts Emerson had dreamed in the first scientific days might be the force that hurled reptile into mammal. I had watched at midnight the mad scientists intent upon their own creation. But in the end, those fantastic flashes of the lightning had ceased without issue, at least for me. The pane, the inscrutable pane, had darkened at last; the scientists, if scientists they were, had departed, carrying their secret with them. I sighed, remembering. It was then I saw the miracle. I saw it because I was hunched at ground level, smelling rank of fox, and no longer gazing with upright human arrogance upon the things of this world.

I did not realize at first what it was that I looked upon. As my wandering attention centered, I saw nothing but two small projecting ears lit by the morning sun. Beneath them, a small neat face looked shyly up at me. The ears moved at every sound, drank in a gull's cry and the far horn of a ship. They crinkled, I began to realize, only with curiosity; they had not learned to fear. The creature was very young. He was alone in a dread universe. I crept on my knees around the prow and crouched beside him. It was a small fox pup from a den under the timbers who looked up at me. God knows what had become of his brothers and sisters. His parent must not have been home from hunting.

He innocently selected what I think was a chicken bone from an untidy pile of splintered rubbish and shook it at me invitingly. There was a vast and playful humor in his face. "If there was only one fox in the world and I could kill him, I would do." The words of a British poacher in a pub rasped in my ears. I dropped even further and painfully away from human stature. It has been said repeatedly that one can never, try as he will, get around to the front of the universe. Man is destined to see only its far side, to realize nature only in retreat.

Yet here was the thing in the midst of the bones, the wide-eyed, innocent fox inviting me to play, with the innate courtesy of its two forepaws placed appealingly together, along with a mock shake of the head. The universe was swinging in some fantastic fashion around to present its face, and the face was so small that the universe itself was laughing.

It was not a time for human dignity. It was a time only for the careful observance of amenities written behind the stars. Gravely I arranged my forepaws while the puppy whimpered with ill-concealed excitement. I drew the breath of a fox's den into my nostrils. On impulse, I picked up clumsily a whiter bone and shook it in teeth that had not entirely forgotten their original purpose. Round and round we tumbled for one ecstatic moment. We were the innocent thing in the midst of the bones, born in the egg, born in the den, born in the dark cave with the stone ax close to hand, born at last in human guise to grow coldly remote in the room with the rifle rack upon the wall.

But I had seen my miracle. I had seen the universe as it begins for all things. It was, in reality, a child's universe, a tiny and laughing universe. I rolled the pup on his back and ran, literally ran for the nearest ridge. The sun was half out of the sea, and the world was swinging back to normal. The adult foxes would be already trotting home.

A little farther on, I passed one on a ridge who knew well I had no gun, for it swung by quite close, stepping delicately with brush and head held high. Its face was watchful but averted. It did not matter. It was what

I had experienced and the fox had experienced, what we had all experienced in adulthood. We passed carefully on our separate ways into the morning, eyes not meeting.

But to me the mist had come, and the mere chance of two lifted sunlit ears at morning. I knew at last why the man on the bed had smiled finally before he dropped his hands. He, too, had worked around to the front of things in his death agony. The hands were playthings and had to be cast aside at last like a little cherished toy. There was a meaning and there was not a meaning, and therein lay the agony.

The meaning was all in the beginning, as though time was awry. It was a little beautiful meaning that did not stay, and the sixty-year-old man on the hospital bed had traveled briefly toward it through the dark at the end of the universe. There was something in the desperate nature of the world that had to be reversed, but he had been too weak to tell me, and the hands had dropped helplessly away.

After forty years I had been just his own age when the fog had come groping for my face. I think I can safely put it down that I had been allowed my miracle. It was very small, as is the way of great things. I had been permitted to correct time's arrow for a space of perhaps five minutes—and that is a boon not granted to all men. If I were to render a report upon this episode, I would say that men must find a way to run the arrow backward. Doubtless it is impossible in the physical world, but in the memory and the will man might achieve the deed if he would try.

For just a moment I had held the universe at bay by the simple expedient of sitting on my haunches before a fox den and tumbling about with a chicken bone. It is the gravest, most meaningful act I shall ever accomplish, but, as Thoreau once remarked of some peculiar errand of his own, there is no use reporting it to the Royal Society.

TEN

The Last Neanderthal

For thou shalt be in league with the stones of the field: and the beasts of the field shall be at peace with thee. —JOB 5:23

IT has long been the thought of science, particularly in evolutionary biology, that nature does not make extended leaps, that her creatures slip in slow disguise from one shape to another. A simple observation will reveal, however, that there are rocks in deserts that glow with heat for a time after sundown. Similar emanations may come from the writer or the scientist. The creative individual is someone upon whom mysterious rays have converged and are again reflected, not necessarily immediately, but in the course of years. That all of this wispy geometry of dreams and memories should be the product of a kind of slow-burning oxidation carried on in an equally diffuse and mediating web of nerve and sense cells is surprising enough, but that the emanations from the same particulate organ, the brain, should be so strikingly different as to disobey the old truism of an unleaping nature is quite surprising, once one comes to contemplate the reality.

The same incident may stand as a simple fact to some, an intangible hint of the nature of the universe to others, a useful myth to a savage, or any number of other things. The receptive mind makes all the difference, shadowing or lighting the original object. I was an observer, intent upon my own solitary hieroglyphics.

It happened a long time ago at Curaçao, in the Netherlands Antilles, on a shore marked by the exposed ribs of a wrecked freighter. The place was one where only a student of desolation would find cause to linger. Pelicans perched awkwardly on what remained of a rusted prow. On the edge of the littered beach beyond the port I had come upon a dead dog wrapped in burlap, obviously buried at sea and drifted in by the waves. The dog was little more than a skeleton but still articulated, one delicate bony paw laid gracefully—as though its owner merely slept, and would presently awaken—across a stone at the water's edge. Around his throat was a waterlogged black strap that showed he had once belonged to someone. This dog was a mongrel whose life had been spent among the island fishermen. He had known only the small sea-beaten boats that come across the strait from Venezuela. He had romped briefly on shores like this to which he had been returned by the indifferent sea.

I stepped back a little hesitantly from the smell of death, but still I paused reluctantly. Why, in this cove littered with tin cans, bottles, and cast-off garments, did I find it difficult, if not a sacrilege, to turn away? Because, the thought finally came to me, this particular tattered garment had once lived. Scenes on the living sea that would never in all eternity recur again had streamed through the sockets of those vanished eyes. The dog was young, the teeth in its jaws still perfect. It was of that type of loving creature who had gamboled happily about the legs of men and striven to partake of their endeavors.

Someone had seen crudely to his sea burial, but not well enough to prevent his lying now where came everything abandoned. Nevertheless, vast natural forces had intervened to clothe him with a pathetic dignity. The tide had brought him quietly at night and placed what remained of him asleep upon the stones. Here at sunrise I had stood above him in a light he would never any longer see. Even if I had had a shovel the stones would have prevented his burial. He would wait for a second tide to spirit him away or lay him higher to bleach starkly upon coral and conch shells, mingling the little lime of his bones with all else that had once stood upright on these shores.

As I turned upward into the hills beyond the beach I was faintly aware of a tracery of lizard tails amidst the sand and the semidesert shrubbery. The lizards were so numerous on the desert floor that their swift movement in the bright sun left a dizzying impression, like spots dancing before one's eyes. The creatures had a tangential way of darting off to the side like inconsequential thoughts that never paused long enough to be fully apprehended. One's eyesight was oppressed by subtly moving points where all should have been quiet. Similar darting specks seemed to be invading my mind. Offshore I could hear the sea wheezing and suspiring in long gasps among the caverns of the coral. The equatorial sun blazed on my unprotected head and hummingbirds flashed like little green flames in the underbrush. I sought quick shelter under a manzanillo tree, whose poisoned apples had tempted the sailors of Columbus.

I suppose the apples really made the connection. Or perhaps merely the interior rustling of the lizards, as I passed some cardboard boxes beside a fence, brought the thing to mind. Or again, it may have been the tropic sun, lending its flames to life with a kind of dreadful indifference as to the result. At any rate, as I shielded my head under the leaves of the poison tree, the darting lizard points began to run together into a pattern.

Before me passed a broken old horse plodding before a cart laden with bags of cast-off clothing, discarded furniture, and abandoned metal. The horse's harness was a makeshift combination of straps mended with rope. The bearded man perched high in the driver's seat looked as though he had been compounded from the assorted junk heap in the wagon bed. What finally occupied the center of my attention, however, was a street sign and a year—a year that scurried into shape with the flickering alacrity of the lizards. "R Street," they spelled, and the year was 1923.

By now the man on the wagon is dead, his cargo dispersed, never to be reassembled. The plodding beast has been overtaken by whatever fate comes upon a junkman's horse. Their significance upon that particular day in 1923 had been resolved to this, just this: The wagon had been passing the intersection between R and Fourteenth streets when I had leaned from a high-school window a block away, absorbed as only a sixteen-year-old may sometimes be with the sudden discovery of time. It is all going, I thought with the bitter desperation of the young confronting history. No one can hold us. Each and all, we are riding into the dark. Even living, we cannot remember half the events of our own days.

At that moment my eye had fallen upon the junk dealer passing his fateful corner. Now, I had thought instantly, now, save him, immortalize the unseizable moment. The junkman is the symbol of all that is going or is gone. He is passing the intersection into nothingness. Say to the mind, "Hold him, do not forget."

The darting lizard points beyond the manzanillo tree converged and tightened. The phantom horse and the heaped, chaotic wagon were still jouncing across the intersection upon R Street. They had never crossed it; they would not. Forty-five years had fled away. I was not wrong about the powers latent in the brain. The scene was still in process.

I estimated the lowering of the sun with one eye while at the back of my mind the lizard rustling continued. The blistering apples of the manzanillo reminded me of an inconsequential wild-plum fall far away in Nebraska. They were not edible but they contained the same, if a simpler, version of the mystery hidden in our heads. They were hoarding and dispersing energy while the inanimate universe was running down around us.

"We must regard the organism as a configuration contrived to evade the tendency of the universal laws of nature," John Joly the geologist had once remarked. Unlike the fire in a thicket, life burned cunningly and hoarded its resources. Energy provisions in the seed provided against individual death. Of all the unexpected qualities of an unexpected universe, the sheer organizing power of animal and plant metabolism is one of the most remarkable, but, as in the case of most everyday marvels, we take it for granted. Where it reaches its highest development, in the human mind, we forget it completely. Yet out of it history is made—the junkman on R Street is prevented from departing. Growing increasingly archaic, that phantom would be held at the R Street intersection while all around him new houses arose and the years passed unremembered. He would not be released until my own mind began to crumble.

The power to free him is not mine. He is held enchanted because long ago I willed a miniature of history, confined to a single brain. That brain is devouring oxygen at a rate out of all proportion to the rest of the body. It is involved in burning, evoking, and transposing visions, whether of lizard tails, alphabets from the sea, or the realms beyond the galaxy. So important does nature regard this unseen combustion, this smoke of the planet's autumn, that a starving man's brain will be protected to the last while his body is steadily consumed. It is a part of unexpected nature.

In the rational universe of the physical laboratory this sullen and obstinate burning might not, save for our habit of taking the existent for granted, have been expected. Nonetheless, it is here, and man is its most tremendous manifestation. One might ask, Would it be possible to understand humanity a little better if one could follow along just a step of the evolutionary pathway in person? Suppose that there still lived . . . but let me tell the tale, make of it what you will.

2

Years after the experience of which I am about to speak, I came upon a recent but Neanderthaloid skull in the dissecting room—a rare enough occurrence, one that the far-out flitting of forgotten genes struggles occasionally to produce, as if life sometimes hesitated and were inclined to turn back upon its pathway. For a time, remembering an episode of my youth, I kept the indices of cranial measurement by me.

Today, thinking of that experience, I have searched vainly for my old notebook. It is gone. The years have a way of caring for things that do not seek the safety of print. The earlier event remains, however, because it was not a matter of measurements or anthropological indices but of a living person whom I once knew.

Now, in my autumn, the face of that girl and the strange season I spent in her neighborhood return in a kind of hazy lesson that I was too young to understand.

It happened in the West, somewhere in that wide drought-ridden land of empty coulees that carry in sudden spates of flood the boulders of the Rockies toward the sea. I suppose that, with the outward flight of population, the region is as wild now as it was then, some forty years ago. It would be useless to search for the place upon a map, though I have tried. Too many years and too many uncertain miles lie behind all bone hunters. There was no town to fix upon a road map. There was only a sod house tucked behind a butte, out of the prevailing wind. And there was a little spring-fed pond in a grassy meadow—that I remember.

Bone hunting is not really a very romantic occupation. One walks day after day along miles of frequently unrewarding outcrop. One grows browner, leaner, and tougher, it is true, but one is far from the bright lights, and the prospect, barring a big strike, like a mammoth, is always to abandon camp and go on. It was really a gypsy profession, then, for those who did the field collecting.

In this case, we did not go on. There was an eroding hill in the vicinity, and on top of that hill, just below sod cover, were the foot bones, hundreds of them, of some lost Tertiary species of American rhinoceros. It is useless to ask why we found only foot bones or why we gathered the mineralized things in such fantastic quantities that they must still lie stacked in some museum storeroom. Maybe the creatures had been immured standing up in a waterhole and in the millions of succeeding years the rest of the carcasses had eroded away from the hilltop stratum. But there were the foot bones, and the orders had come down, so we dug carpals and metacarpals till we cursed like an army platoon that headquarters has forgotten.

There was just one diversion: the spring, and the pond in the meadow. There, under the bank, we cooled our milk and butter purchased from the soddy inhabitants. There we swam and splashed after work. The country people were reserved and kept mostly to themselves. They were uninterested in the dull bones on the hilltop unenlivened by skulls or treasure. After all, there was reason for their reserve. We must have appeared, by their rural standards, harmless but undoubtedly touched in the head. The barrier of reserve was never broken. The surly farmer kept to his parched acres and estimated to his profit our damage to his uncultivated hilltop. The slatternly wife tended a few scrawny chickens. In that ever blowing landscape their windmill largely ran itself.

Only a stocky barefoot girl of twenty sometimes came hesitantly down the path to our camp to deliver eggs. Some sixty days had drifted by upon that hillside. I began to remember the remark of an old fossil hunter who in his time had known the Gold Coast and the African veldt. "When calico begins to look like silk," he had once warned over a fire in the Sierras, "it is time to go home."

But enough of that. We were not bad young people. The girl shyly brought us the eggs, the butter, and the bacon, and then withdrew. Only after some little time did her appearance begin to strike me as odd. Men are accustomed to men in their various color variations around the world. When the past intrudes into a modern setting, however, it is less apt to be visible, because to see it demands knowledge of the past, and the past is always camouflaged when it wears the clothes of the present.

The girl came slowly down the trail one evening, and it struck me suddenly how alone she looked and how, well, *alien,* she also appeared. Our cook was stoking up the evening fire, and as the shadows leaped and flickered I, leaning invisibly against a rock, was suddenly transported one hundred thousand years into the past. The shadows and their dancing highlights were the cause of it. They had swept the present out of sight. That girl coming reluctantly down the pathway to the fire was removed from us in time, and subconsciously she knew it as I did. By modern standards she was not pretty, and the gingham dress she wore, if anything, defined the difference.

Short, thickset, and massive, her body was still not the body of a typical peasant woman. Her head, thrust a little forward against the light, was massive-boned. Along the eye orbits at the edge of the frontal bone I could see outlined in the flames an armored protuberance that, particularly in women, had vanished before the close of the Würmian ice. She swung her head almost like a great muzzle beneath its curls, and I was struck by the low bun-shaped breadth at the back.

Along her exposed arms one could see a flash of golden hair.

No, we are out of time, I thought quickly. We are each and every one displaced. She is the last Neanderthal, and she does not know what to do. We are those who eliminated her long ago. It is like an old scene endlessly re-enacted. Only the chipped stones and the dead game are lacking.

I came out of the shadow then and spoke gently to her, taking the packages. It was the most one could do across that waste of infinite years. She spoke almost inaudibly, drawing an unconscious circle in the dust with a splayed bare foot. I saw, through the thin dress, the powerful thighs, the yearning fertility going unmated in this lonesome spot. She looked up, and a trick of the fire accentuated the cavernous eye sockets so that I saw only darkness within. I accompanied her a short distance along the trail. "What is it you are digging for?" she managed to ask.

"It has to do with time," I said slowly. "Something that happened a long time ago."

She listened incuriously, as one at the morning of creation might do.

"Do you like this?" she persisted. "Do you always just go from one place to another digging these things? And who pays for it, and what comes of it in the end? Do you have a home?" The soddy and her burly father were looming in the dusk. I paused, but questions flung across the centuries are hard to answer.

"I am a student," I said, but with no confidence.

How could I say that suddenly she herself and her ulnar-bowed and golden-haired forearms were a part of a long reach backward into time?

"Of what has been, and what will come of it we are trying to find out. I am afraid it will not help one to find a home," I said, more to myself than her. "Quite the reverse, in fact. You see—"

The dark sockets under the tumbled hair seemed somehow sadly vacant. "Thank you for bringing the things," I said, knowing the customs of that land. "Your father is waiting. I will go back to camp now." With this I strode off toward our fire but went, on impulse, beyond it into the full-starred night.

This was the way of things along the Wild Cat escarpment. There was sand blowing and the past mingling with the present in more ways than professional science chose to see. There were eroded farms no longer running cattle and a diminishing population waiting, as this girl was waiting, for something they would never possess. They were, without realizing it, huntsmen without game, women without warriors. Obsolescence was upon their way of life.

But about the girl lingered a curious gentleness that we know now had long ago touched the vanished Neanderthals she so resembled. It would be her fate to marry eventually one of the illiterate hard-eyed uplanders of my own kind. Whatever the subtle genes had resurrected in her body would be buried once more and hidden in the creature called *sapiens*. Perhaps in the end his last woman would stand unwanted before some fiercer, brighter version of himself. It would be no more than justice. I was farther out in the deep spaces than I knew, and the fire was embers when I returned.

The season was waning. There came, inevitably, a time when the trees began to talk of winter in the crags above the camp. I have repeated all that can be said about so fragile an episode. I had exchanged in the course of weeks a few wistful, scarcely understood remarks. I had waved to her a time or so from the quarry hilltop. As the time of our departure neared I had once glimpsed her shyly surveying from a rise beyond the pond our youthful plungings and naked wallowings in the spring-fed water. Then suddenly the leaves were down or turning yellow. It was time to go. The fossil quarry and its interminable foot bones were at last exhausted.

But something never intended had arisen for me there by the darkening water—some agonizing, lifelong nostalgia, both personal and, in another sense, transcending the personal. It was—how shall I say it?—the endurance in a single mind of two stages of man's climb up the energy ladder that may be both his triumph and his doom.

Our battered equipment was assembled in the Model T's, which, in that time, were the only penetrators of deep-rutted upland roads. Morose good-byes were expressed; money was passed over the broken sod cover on the hilltop. Hundreds of once galloping rhinoceros foot bones were stowed safely away. And that was it.

I stood by the running board and slowly, very slowly, let my eyes wander toward that massive, archaic, and yet tragically noble head—of a creature so far back in time it did not know it represented tragedy. I made, I think, some kind of little personal gesture of farewell. Her head raised in recognition and then dropped. The motors started. *Homo sapiens,* the energy devourer, was on his way once more.

What was it she had said, I thought desperately as I swung aboard. Home, she had questioned, "Do you have a home?" Perhaps I once did, I was to think many times in the years that followed, but I, too, was a mental atavism. I, like that lost creature, would never find the place called home. It lay somewhere in the past down that hundred-thousand-year road on which travel was impossible. Only ghosts with uncertain eyes and abashed gestures would meet there. Upon a surging tide of power first conceived in the hearth fires of dead caverns mankind was plunging into an uncontrolled future beyond anything the people of the Ice had known.

The cell that had somehow mastered the secret of controlled energy, of surreptitious burning to a purpose, had finally produced the mind, judiciously, in its turn, controlling the inconstant fire at the cave mouth. Beyond anything that lost girl could imagine, words in the mouth or immured in libraries would cause substance to vanish and the earth itself to tremble. The little increments of individual energy dissolving at death had been coded and passed through the centuries by human ingenuity. A climbing juggernaut of power was leaping from triumph to triumph. It threatened to be more than man and all his words could master. It was more and less than man.

I remembered those cavernous eye sockets whose depths were forever hidden from me in the firelight. Did they contain a premonition of the end we had invited, or was it only that I was young and hungry for all that was untouchable? I have searched once more for the old notebooks but, again, in vain. They would tell me, at best, only how living phantoms can be anatomically compared with those of the past. They would tell nothing of that season of the falling leaves or how I learned under the night sky of the utter homelessness of man.

3

I have seen a tree root burst a rock face on a mountain or slowly wrench aside the gateway of a forgotten city. This is a very cunning feat, which men take too readily for granted. Life, unlike the inanimate, will take the long way round to circumvent barrenness. A kind of desperate will resides even in a root. It will perform the evasive tactics of an army, slowly inching its way through crevices and hoarding energy until someday it swells and a living tree upheaves the heaviest mausoleum. This covert struggle is part of the lifelong battle waged against the Second Law of Thermodynamics, the heat death that has been frequently assumed to rule the universe. At the hands of man that hoarded energy takes strange forms, both in the methods of its accumulation and in the diverse ways of its expenditure.

For hundreds of thousands of years, a time longer than all we know of recorded history, the kin of that phantom girl had lived without cities along the Italian Mediterranean or below the northern tentacles of the groping ice. The low archaic skull vault had been as capacious as ours. Neanderthal man had, we now know after long digging, his own small dreams and kindnesses. He had buried his dead with offerings—there were even evidences that they had been laid, in some instances, upon beds of wild flowers. Beyond the chipped flints and the fires in the cavern darkness his mind had not involved itself with what was to come upon him with our kind—the first bowmen, the great artists, the terrible creatures of his blood who were never still.

It was a time of autumn driftage that might have lasted and been well forever. Whether it was his own heavy brow that changed in the chill nights or that somewhere his line had mingled with a changeling cuckoo brood who multiplied at his expense we do not know with certainty. We know only that he vanished, though sometimes, as in the case of my upland girl, a chance assemblage of archaic genes struggles to reemerge from the loins of *sapiens*.

But the plucked flint had flown; the heavy sad girls had borne the children of the conquerors. Rain and leaves washed over the cave shelters of the past. Bronze replaced flint, iron replaced bronze, while the killing never ceased. The Neanderthals were forgotten; their grottoes housed the oracles of later religions. Marble cities gleamed along the Mediterranean. The ice and the cave bear had vanished. White-robed philosophers discoursed in Athens. Armed galleys moved upon the waters. Agriculture had brought wealth and diversification of labor. It had also brought professional soldiery. The armored ones were growing and, with them, slavery, torture, and death upon all the seas of the world.

The energy that had once sufficed only to take man from one camping place to another, the harsh but innocent world glimpsed by Cook in the eighteenth century on the shores of Australia, century by century was driving toward a climax. The warriors with the tall foreheads given increasingly to fanatic religions and monumental art had finally grown to doubt the creations of their own minds.

The remnants of what had once been talked about in Athens and been consumed in the flames of Alexandria hesitantly crept forth once more. Early in the seventeenth century Sir Francis Bacon asserted that "by the agency of man a new aspect of things, a new universe, comes into view." In those words he was laying the basis of what he came to call "the second world," that world which could be drawn out of the natural by the sheer power of the human mind. Man had, of course, unwittingly been doing something of the sort since he came to speak. Bacon, however, was dreaming of the new world of invention, of toleration, of escape from irrational custom. He was the herald of the scientific method itself. Yet that method demands history also—the history I as an eager student had long ago beheld symbolically upon a corner in the shape of a junkman's cart. Without knowledge of the past, the way into the thickets of the future is desperate and unclear.

Bacon's second world is now so much with us that it rocks our conception of what the natural order was, or is, or in what sense it can be restored. A mathematical formula traveling weakly along the fibers of the neopallium may serve to wreck the planet. It is a kind of metabolic energy never envisaged by the lichen attacking a rock face or dreamed of in the flickering shadows of a cave fire. Yet from these ancient sources man's hunger has been drawn. Its potential is to be found in the life of the world we call natural, just as its terrifying intricacy is the product of the second visionary world evoked in the brain of man.

The two exist on the planet in an increasingly uneven balance. Into one or the other or into a terrifying nothing one of these two worlds must finally subside. Man, whose strange metabolism has passed beyond the search for food to the naked ingestion of power, is scarcely aware that the energy whose limited planetary store lies at the root of the struggle for existence has passed by way of his mind into another dimension. There the giant shadows of the past continue to contend. They do so because life is a furnace of concealed flame.

Some pages back I spoke of a wild-plum thicket. I did so because I had a youthful memory of visiting it in autumn. All the hoarded juices of summer had fallen with that lush untasted fruit upon the grass. The tiny engines of the plant had painstakingly gathered throughout the summer rich stores of sugar and syrup from the ground. Seed had been produced; birds had flown away with fruit that would give rise to plum trees miles away. The energy dispersion was so beneficent on that autumn afternoon that earth itself seemed anxious to promote the process against the downward guttering of the stars. Even I, tasting the fruit, was in my animal way scooping up some of it into thoughts and dreams.

Long after the Antillean adventure I chanced on an autumn walk to revisit the plum thicket. I was older, much older, and I had come largely because I wondered if the thicket was still there and because this strange hoarding and burning at the heart of life still puzzled me. I have spoken figuratively of fire as an animal, as being perhaps the very *essence* of animal. Oxidation, I mean, as it enters into life and consciousness.

Fire, as we have learned to our cost, has an insatiable hunger to be fed. It is a nonliving force that can even locomote itself. What if now—and I half closed my eyes against the blue plums and the smoke drifting along the draw—what if now it is only concealed and grown slyly conscious of its own burning in this little house of sticks and clay that I inhabit? What if I am, in some way, only a sophisticated fire that has acquired an ability to regulate its rate of combustion and to hoard its fuel in order to see and walk?

The plums, like some gift given from no one to no

one visible, continued to fall about me. I was old now, I thought suddenly, glancing at a vein on my hand. I would have to hoard what remained of the embers. I thought of the junkman's horse and tried to release him so that he might be gone.

Perhaps I had finally succeeded. I do not know. I remembered that star-filled night years ago on the escarpment and the heavy-headed dreaming girl drawing a circle in the dust. Perhaps it was time itself she drew, for my own head was growing heavy and the smoke from the autumn fields seemed to be penetrating my mind. I wanted to drop them at last, these carefully hoarded memories. I wanted to strew them like the blue plums in some gesture of love toward the universe all outward on a mat of leaves. Rich, rich and not to be hoarded, only to be laid down for someone, anyone, no longer to be carried and remembered in pain like the delicate paw lying forever on the beach at Curaçao.

I leaned farther back, relaxing in the leaves. It was a feeling I had never had before, and it was strangely soothing. Perhaps I was no longer *Homo sapiens,* and perhaps that girl, the last Neanderthal, had known as much from the first. Perhaps all I was, really, was a pile of autumn leaves seeing smoke wraiths through the haze of my own burning. Things get odder on this planet, not less so. I dropped my head finally and gazed straight up through the branches at the sun. It was all going, I felt, memories dropping away in that high indifferent blaze that tolerated no other light. I let it be so for a little, but then I felt in my pocket the flint blade that I had carried all those years from the gravels on the escarpment. It reminded me of a journey I would not complete and the circle in the dust around which I had magically traveled for so long.

I arose then and, biting a plum that tasted bitter, I limped off down the ravine. One hundred thousand years had made little difference—at least, to me. The secret was to travel always in the first world, not the second; or, at least, to know at each crossroad which world was which. I went on, clutching for stability the flint knife in my pocket. A blue smoke like some final conflagration swept out of the draw and preceded me. I could feel its heat. I coughed, and my eyes watered. I tried as best I could to keep pace with it as it swirled on. There was a crackling behind me as though I myself were burning, but the smoke was what I followed. I held the sharp flint like a dowser's twig, cold and steady in my hand.

Barlow, Nora, "Charles Darwin and the Galápagos Islands," *Nature,* Vol. 136 (1935), p. 391.

Beagle, Peter, *The Last Unicorn,* New York, Viking Press, 1968.

Beaglehole, J. C., *The Journals of Captain James Cook on His Voyages of Discovery,* 3 vols. to date and a portfolio, London, Cambridge University Press, 1955–67.

Brown, Lloyd A., *The Story of Maps,* New York, Crown Publishers, 1949.

Cameron, Hector Charles, *Sir Joseph Banks: The Autocrat of the Philosophers,* London, The Batchworth Press, 1952.

Campbell, Joseph, *The Hero with a Thousand Faces,* New York, Meridian Books, 1956.

——— *The Masks of God: Primitive Mythology,* New York, Viking Press, 1959.

Carlquist, Sherwin, *Island Life: A Natural History of the Islands of the World,* Garden City, N.Y., The Natural History Press, 1965.

Carozzi, Albert V., "Agassiz's Amazing Geological Speculation: the Ice Age," *Studies in Romanticism,* Vol. 5 (1966), pp. 57–83.

Carrington, Hugh, *Life of Captain Cook,* London, Sidgwick and Jackson, 1939.

Chaning-Pearce, Melville, *The Terrible Crystal,* London, Kegan Paul, 1940.

Channing, William Ellery, *Thoreau, the Poet Naturalist,* New York, Biblo and Tannen, 1966.

Chapman, Walker, *The Loneliest Continent,* Greenwich, Conn., New York Graphic Society Publishers Ltd., 1964.

Charlesworth, J. K., *The Quaternary Era: With Special Reference to Its Glaciation,* 2 vols., London, Edward Arnold, 1957.

Cherry-Garrard, Apsley, *The Worst Journey in the World,* London, Penguin Books, 1948.

Christie, John Aldrich, *Thoreau as World Traveler,* New York, Columbia University Press, 1965.

Clarke, Howard W., *The Art of the Odyssey,* Englewood Cliffs, N.J., Prentice-Hall, 1967.

Coburn, Kathleen (ed.), *Inquiring Spirit: A New Presentation of Coleridge from His Published and Unpublished Writings,* New York, Pantheon Books, 1951.

Creed, John Martin, and Boys Smith, John Sandwith (eds.), *Religious Thought in the Eighteenth Century,* London, Cambridge University Press, 1934.

D'Ancona, Umberto, *The Struggle for Existence,* Leiden, Holland, E. J. Brill, 1954.

Darwin, Charles, *Journal of Researches,* London, Henry Colburn, 1839.

——— *On the Origin of Species,* reprint of 2nd rev. ed., London, Oxford University Press, 1860.

Darwin, Francis (ed.), *Life and Letters of Charles Darwin,* 3 vols., London, John Murray, 1888.

——— and Seward, A. C. (eds.), *More Letters of Charles Darwin,* London, John Murray, 1903.

Dyson, James L., *The World of Ice,* New York, Alfred Knopf, 1962.

Eiseley, Loren, *Francis Bacon and the Modern Dilemma,* Lincoln, University of Nebraska Press, 1963.

——— "Neanderthal Man and the Dawn of Human Paleontology," *The Quarterly Review of Biology,* Vol. 32 (December 1957), pp. 323–329.

Eydoux, Henri-Paul, *The Buried Past: A Survey of Great Archaeological Discoveries,* New York, Frederick A. Praeger, 1966.

Gelli, Giovanni Battista, *Circe,* trans. by Thomas Brown, ed. by Robert M. Adams, Ithaca, N.Y., Cornell University Press, 1963.

Gibbs, Frederic Andrews, "The Most Important Thing," *American Journal of Public Health,* Vol. 41 (1951), pp. 1503–1508.

Harding, Walter (ed.), *Thoreau, Man of Concord,* New York, Holt, Rinehart and Winston, 1960.

Heschel, Abraham J., *Who Is Man?,* Stanford, Stanford University Press, 1965.

Hooker, Sir Joseph Dalton (ed.), *Journal of the Right Honorable Sir Joseph Banks,* London, Macmillan, 1896.

Hyman, Stanley Edgar, "Descent, Fall and Sex—Darwin's Victorianism," *The Carleton Miscellany,* Vol. 2, No. 4 (1961), pp. 11–25.

——— *The Tangled Bank: Darwin, Marx, Frazer and Freud as Imaginative Writers,* New York, Atheneum, 1962.

Ivanova, I. K., "The Significance of Fossil Hominids and Their Culture for the Stratigraphy of the Quarternary Period," *Arctic Anthropology,* Vol. 4 (1967), pp. 212–223.

Jolly, Alison, *Lemur Behavior: A Madagascar Field Study,* Chicago, University of Chicago Press, 1966.

Joly, John, *The Birth-Time of the World, and Other Scientific Essays,* New York, E. P. Dutton, 1915.

Kazantzakis, Nikos, *The Odyssey: A Modern Sequel,* New York, Simon and Schuster, 1958.

Lattimore, Richmond, *The Odyssey of Homer: A Modern Translation,* New York, Harper and Row, 1965.

MacArthur, Robert H., and Wilson, E. O., *The Theory of Island Biogeography,* Princeton, Princeton University Press, 1967.

McGlashan, Alan, *The Savage and Beautiful Country,* London, Chatto and Windus, 1966.

Mayr, Ernst, "Changes of Genetic Environment and Evolution," in *Evolution as a Process,* ed. by Julian Huxley *et al.,* New York, Collier Books, 1963.

Mellersh, H. E. L., *Fitzroy of the Beagle,* London, Rupert Hart-Davis, 1968.

Partridge, R. B., "The Primeval Fireball Today," *The American Scientist,* Vol. 57 (1969), pp. 37–74.

Pascoli, Giovanni, *Poesie,* Verona, Italy, Arnoldo Mondadori Editore, 1965.

Price, A. Grenfell (ed.), *The Explorations of Captain James Cook in the Pacific, as told by Selections of His Own Journals, 1768–1779,* New York, Heritage Press, 1958.

Ritchie, James, "The Edinburgh Explorers," *University of Edinburgh Journal,* Vol. 12 (1943), pp. 155–159.

——— "Evolution and the Galápagos Islands," *University of Edinburgh Journal,* Vol. 12 (1943), pp. 97–105.

Santayana, George, *The Birth of Reason,* New York, Columbia University Press, 1968.

Schultz, Gwen, *Glaciers and the Ice Age,* New York, Holt, Rinehart and Winston, 1963.

Sillman, Leonard, "The Genesis of Man," *International Journal of Psychoanalysis,* Vol. 34 (1953), pp. 146–152.

Simpson, George Gaylord, *The Geography of Evolution,* New York and Philadelphia, Chilton Books, 1965.

Smith, Edward, *The Life of Sir Joseph Banks: President of the Royal Society, with Some Notices of His Friends and Contemporaries,* London, John Lane, The Bodley Head, 1911.

Smith, Joseph Lindon, *Tombs, Temples and Ancient Art,* Norman, Okla., University of Oklahoma Press, 1956.

Stanford, W. B., *The Ulysses Theme: A Study in the Adaptability of a Traditional Hero,* 2nd rev. ed., Oxford, England, Basil Blackwell, 1963.

Sullivan, Walter, "The Neanderthal Man Liked Flowers," *The New York Times,* June 13, 1968.

Thoreau, Henry David, *The Journal of Henry David Thoreau,* ed. by Bradford Torrey and Francis H. Allen, 14 vols., Boston, Houghton Mifflin, 1949.

Villiers, Alan, *Captain James Cook,* New York, Charles Scribner's Sons, 1967.

——— *The Coral Sea,* New York, Whittlesey House, McGraw-Hill, 1949.

Wallace, Alfred Russel, *Island Life,* 2nd rev. ed., London, Macmillan & Co., 1892.

Wheeler, John Archibald, "Our Universe: The Known and the Unknown," *The American Scholar,* Vol. 37 (1968), pp. 248–274.

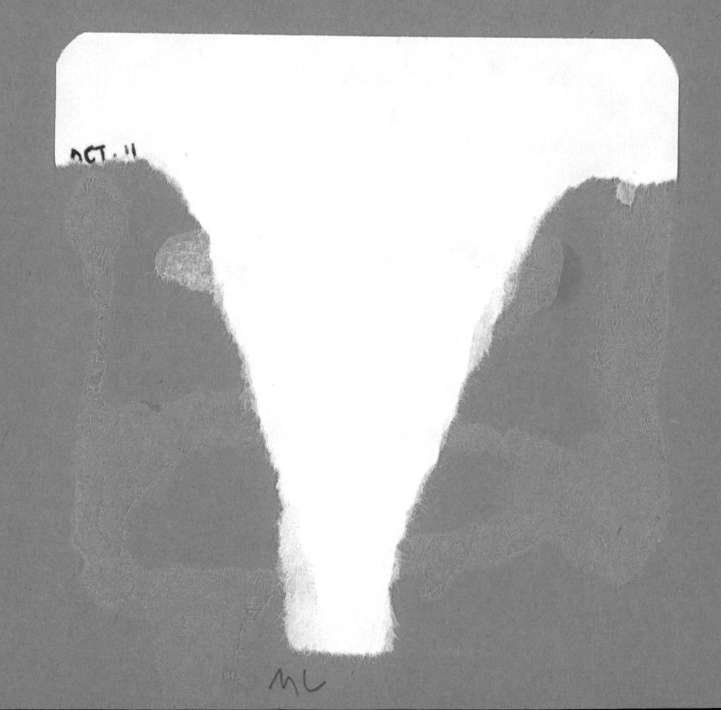